11th International Conference on Manufacturing Science and Technology (ICMST 2020)

Held online due to COVID-19

IOP Conference Series: Materials Science and Engineering
Volume 1063

Liverpool, United Kingdom
22 - 24 September 2020

ISBN: 978-1-7138-3341-3
ISSN: 1757-8981

Printed from e-media with permission by:

Curran Associates, Inc.
57 Morehouse Lane
Red Hook, NY 12571

Some format issues inherent in the e-media version may also appear in this print version.

Institute of Physics
Dirac House, Temple Back
Bristol BS1 6BE UK

Phone: 44 1 17 929 7481
Fax: 44 1 17 920 0979

techtracking@iop.org

Additional copies of this publication are available from:

Curran Associates, Inc.
57 Morehouse Lane
Red Hook, NY 12571 USA
Phone: 845-758-0400
Fax: 845-758-2633
Email: curran@proceedings.com
Web: www.proceedings.com

TABLE OF CONTENTS

Preface

Under original plan, the 11th International Conference on Manufacturing Science and Technology (ICMST 2020) would be held in University of Liverpool, Liverpool, UK during September 22-24, 2020. However, due to the global COVID-19 pandemic, ICMST 2020 was broadcasted through web platform (ZOOM) and successfully held during September 22-24, 2020.

With the great efforts and generous support of conference committees, reviewers, and session chairs, as well as all participants, presenters, and the main organisers - Chemistry and Materials Society (CMS) group from Hong Kong, China, ICMST 2020 was a complete success. The access to ICMST 2020 web platform was offered to all of attendees, which connects all of them from all over the world. All sessions, oral presentations, discussions, and other activities were offered thought ICMST 2020 web platform.

The conference consists of 5 keynote speeches and 2 parallel sessions (13 participants in total). Keynote speakers were from USA, Portugal. Each of them delivered wonderful speeches about 40 minutes, including 35 for presentation and 5 minutes for Q&A. And the 13 participants of 2 parallel sessions who presented their work were from China, Iraq, Saudi Arabia, France, USA, Thailand and so on. Each presentation lasts for 15 minutes, including 12 minutes for presentation and 3 minutes for Q&A. Even though it's online platform, all of them still actively exchanged thoughts, shared opinion, etc.

All participants were immersed in the conference. They listened to the presentations attentively, put forward their opinion. And presenters answered questions actively. The entire atmosphere was very well. To make sure that the conference went more smoothly and all participants can present their work without any problem, the main organisers provided a test session for all presenters in advance. They did their best to help all participants to deal with any problems that they encountered. Meanwhile, video presentations were acceptable, which was for those authors that had internet connection problem (unstable or weak internet connection). And the whole conference was recorded for backup purposes only.

The ICMST aims to provide a platform to help the researchers, scholars and so on communicate their thought, idea, innovation and so on about the area of Manufacturing Science and Technology. The conference proceeding was composed of 11 papers covering nanomaterials and nanomanufacturing, Rapid manufacturing technologies, etc. We sincerely hope that you will find the ICMST conference proceeding beneficial and fruitful for your professional development.

Prof. Yuyuan Zhao
University of Liverpool, UK

IOP Conf. Series: Materials Science and Engineering 1063 (2021) 011001 doi:10.1088/1757-899X/1063/1/011001

Conference Committee

Conference Chairs
Prof. Yuyuan Zhao, University of Liverpool, UK
Prof. Jesús Toribio, University of Salamanca, Spain

Program Chairs
Prof. Sujan Debnath, Curtin University, Malaysia
Assoc. Prof. Jamaludin Jalani, UniversitiTun Hussein Onn Malaysia, Malaysia

Technical Committees
S. Joseph Antony, University of Leeds, UK
François Ducobu, University of Mons, Belgium
Philippe Boisse, LamCoS, France
Lin Mingxing, Shandong University, China
GuohuaCui, Hebei University of Engineering, China
Pei-Hsing Huang, National Yunlin University of Science and Technology, Taiwan
En-Chih Chang, I-Shou University, Taiwan
Jee Hou Ho, The University of Nottingham Malaysia Campus, Malaysia
Yap Hwa Jen , University of Malaya, Malaysia
Baharudin B.T.H.T,Universiti Putra Malaysia, Malaysia
Ahmad Razlan Bin Yusoff, Universiti Malaysia Pahang, Malaysia
Abu Bakar bin Ibrahim, Sultan Idris Education University, Malaysia
Lola Domnina Pestaño, University of Santo Tomas, Philippines
K. S. Vijay Sekar, SSN College of Engineering, India
Pisut Koomsap, Asian Institute of Technology, Thailand
Sarum Boonmee, Suranaree University of Technology, Thailand
Anshuman Srivastava, Mechanical Engg Dept, SIET, Allahabad, India
Mohandas K. N., Ramaiah Institute of Technology, India
Badari Narayana Kantheti, Ramaiah Institute of Technology, India
Anilkumar T., Ramaiah Institute of Technology, India
P. L. Srinivasa Murthy, Ramaiah Institute of Technology, India
V. Anandakrishnan, National Institute of Technology, India
K. Panneerselvam, National Institute of Technology, India
N. V. Raghavendra, National Institute of Engineering, Mysuru, India
Ranganath M. S, Delhi Technological University, India
D. Lenin Singaravelu, National Institute of Technology, India
Md. Anayet U Patwari, Islamic University of Technology, Bangladesh
Mohammad Ahsan Habib, Islamic University of Technology, Bangladesh
Md. Shafiul Ferdous, Khulna University of Engineering & Technology, Bangladesh
Mahmoud Baniasadi, Georgia Southern University, USA

Reviewer Committees
Yang Xi, Newomics Inc., USAVladimir Gurau, Georgia Southern University, USA
Yue Zhang, Georgia Southern University, USA

IOP Conf. Series: Materials Science and Engineering 1063 (2021) 011001 doi:10.1088/1757-899X/1063/1/011001

Zhe Xu, Thayer School of Engineering Dartmouth College, USA
Paulo Nobre Balbis dos Reis, University of Beira Interior, Portugal
Radu Godina, NOVA University Lisbon, Portugal
Suntree Sangjan, Kasetsart University, Kamphaeng Saen campus, Thailand
Yusri Bin Hj Yusof, Universiti Tun Hussein Onn Malaysia (UTHM), Malaysia
Wan Hasrulnizzam Wan Mahmood, Universiti Teknikal Malaysia Melaka, Malaysia
Mahmood Anwar, Curtin University, Sarawak, Malaysia
Mohd Hasmizam Razali, Universiti Malaysia Terengganu, Malaysia
Che Zalina binti Zulkifli, Sultan Idris Education University, Malaysia
Anika Zafiah Mohd Rus, Universiti Tun Hussien Onn Malaysia, Malaysia
Raghed ibrahim Esmaeel, Universiti Teknologi Malaysia, Malaysia
Lydia Anggraini, President University, Indonesia
Chih-Ling Huang, Kaohsiung Medical University, Taiwan
John M. Ikome, Vaal University of Technology, South Africa
Asma Perveen, Nazaryabev University, Kazakhistan
Chang-Seon Shon, Nazarbayev University, Kazakhstan
Muhammad Azhar Ali Khan, Prince Mohammad Bin Fahd University, Saudi Arabia
T.Theivasanthi, International Research Center, Kalasalingam University, India
Vishwanth Koti, Ramaiah Institute of Technology, India
Hemavathy S, Ramaiah Institute of Technology, India
Prakrathi S., Ramaiah Institute of Technology, India
M R Sanjay, Ramaiah Institute of Technology, India

Peer review declaration

All papers published in this volume of *IOP Conference Series: Materials Science and Engineering* have been peer reviewed through processes administered by the Editors. Reviews were conducted by expert referees to the professional and scientific standards expected of a proceedings journal published by IOP Publishing.

- **Type of peer review:** Double-blind
- **Conference submission management system:** iConf Conference System
- **Number of submissions received:** 25
- **Number of submissions sent for review:** 25
- **Number of submissions accepted:** 13
- **Acceptance Rate (Number of Submissions Accepted / Number of Submissions Received X 100):** 52%
- **Average number of reviews per paper:** 3
- **Total number of reviewers involved:** 16
- **Any additional info on review process:**

All received submissions were subjected to double-blind peer reviews, who are expert or have been experiencing in the related field for years. The accepted papers must be revised according to the referees' comments and suggestions, before inclusion in the conference proceedings. Manuscripts submitted should be original & unpublished and written in accordance with the standard IOP paper template.

- **Contact person for queries:** Ms. Jane Li (icmst@cbees.net)

Experimental and simulation requirements for residual stress of TC4 titanium alloy based on ultrasonic rolling

Yanjie Liu[1], Xiuli Fu[1], Hongxia Li[2], Pengcheng Wang[1], Xiuhua Men[1*]

[1]School of Mechanical Engineering, University of Jinan, Jinan, CO 250022, China
[2]Shandong Huayun Electrical and Mechanical Technology Co., Ltd, Jinan, CO 250101 China

*E-mail: me_menxh@ujn.edu.cn

Abstract. The residual compressive stress can improve the fatigue performance and strengthen the material surface. In this paper, TC4 titanium alloy is chosen as the material to measure surface residual stress under different static pressures by automatic gradient stress detection and analysis system. ABAQUS software is used to simulate the variation of residual stress in ultrasonic rolling process. The results reveal that the experimental values and the simulation values are consistent basically. The residual compressive stress reaches its maximum when the static pressure is 1000N. When the static pressure is 1100N ~ 1200N, the depth of the residual compressive stress layer reaches its maximum.

1. Introduction

The titanium alloy is used extensively in the field of aeroengine[1,2]. Fatigue failure becomes the main failure mode of titanium alloy components in this complex service environment of high temperature and high pressure. At present, the surface damage and surface layer residual tensile stress caused by forming and manufacturing process have become the most prominent problems which destroy the fatigue performance of aviation key components. However, the residual compressive stress can restrain the generation and development of the surface cracks, thus improving the fatigue performance[3]. Therefore, the formation of surface layer residual stress has been a hot topic in recent years. With the development of research and application of ultrasonic vibration technology, the ultrasonic rolling shows many advantages as a new surface strengthening process[4]. Ultrasonic rolling device is simple and it can be directly installed on a lathe or milling machine as an independent machining device. Its processing path is controllable. The flowing micro-plastic deformation on the surface layer can be produced by ultrasonic rolling. At the same time, the residual compressive stress on the surface layer increases obviously[5].

In recent years, the influence of ultrasonic rolling parameters on surface layer residual stress has been extensively studied. Zheng Jianxin et al.[6] simulated the ultrasonic rolling of 7075 aluminum alloy using ANSYS software. The reasults indicated that the depth of residual compressive stress layer was 1.05mm. The maximum residual compressive stress reached 285MPa when it was 0.31mm away from the surface. Li Fengqin et al.[7] carried out ultrasonic rolling tests on TC4 titanium alloy under different static pressures with CA6140 lathe and tested the residual stress through experiments. The results showed that the maximum residual stress was about 260MPa when the static pressure was 600N. Liu Zhihua et al.[8] measured the surface residual stress of 18CrNiMo7-6 gear steel. It was found that the maximum residual compressive stress of surface layer increased by 667MPa under the effect of ultrasonic rolling. And the depth of residual compressive stress layer increased by 660 μm.

By analyzing the researches above, it can be seen that ultrasonic rolling contributes to form a certain depth of residual compressive stress layer. In addition, the residual stress can be measured by finite element simulation or measurement experiment. In this paper, the surface layer residual stress of the sample is measured by automatic gradient stress detection and analysis system. Then the residual stress is simulated by finite element method using ABAQUS software, which improves the reliability of the results.

2. Experimental procedures

2.1. Material and sample preparation

In the ultrasonic rolling experiment, TC4 titanium alloy plate is used as the material and cut into 25mm × 50mm × 25mm sample. Its chemical composition is listed in table 1. The surface quality of specimens after ultrasonic rolling is affected by parameters such as amplitude, static pressure, feed rate and rolling times. Therefore, the single-factor test method is usually used for experimental researches. H+VM1260 type ultrasonic rolling device was used for surface machining of specimens. The specific process parameters are listed in table 2.

Table 1. Main chemical composition of TC4 titanium alloy (wt%)

Elements	Al	V	Fe	O	C	N	H	Ti
Proportions	5.5-6.8	3.5-4.5	≤0.3	≤0.2	≤0.1	≤0.05	≤0.015	The rest

Table 2. Process parameters in ultrasonic rolling process

Parameters	Static pressure (N)	Amplitude (μm)	Feed rate (m/min)	Frequency (KHz)	Rolling times
Numerical values	900/1000/1100/1200	7	2	27	1

2.2. Residual stress measurement

Blind hole method was first proposed by Mathar J in 1934. Nowadays it has been developed more mature and is widely used in residual stress detection. The automatic gradient stress detection and analysis system (as shown in figure 1) is used to measure the residual stress of the sample combined with the principle of blind hole method. Paste the strain gauge at the test point, and drill a certain depth of the test point. The drilling position and the paste method of the strain gauge are displayed in figure 2. It will cause the change of resistance of the strain gauge when the stress of the borehole is released. Then the corresponding residual stress can be obtained by the residual stress analysis system.

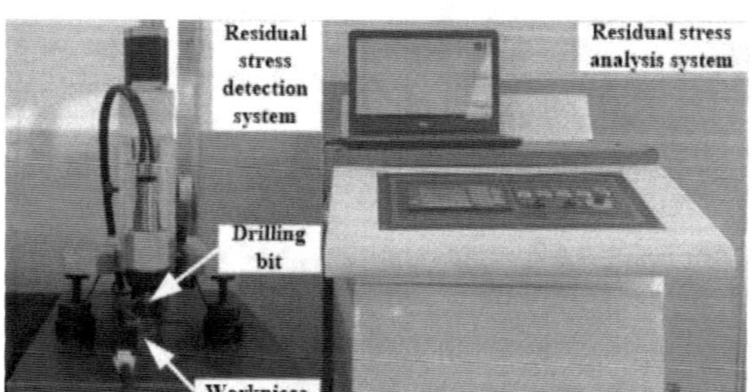

Figure 1. Automatic gradient stress detection and analysis system.

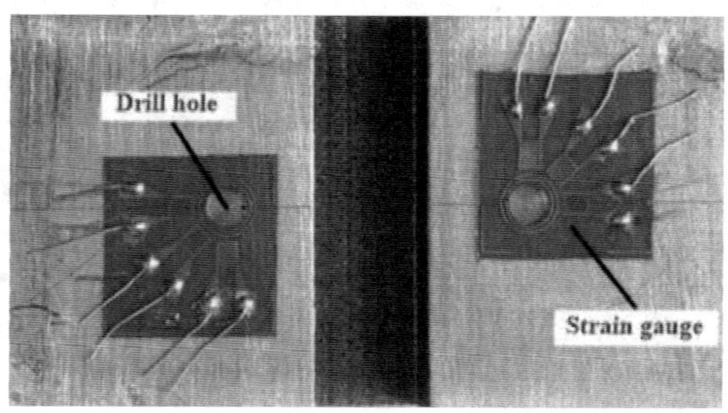

Figure 2. Paste method of the strain gauge.

2.3. Finite element model

In the finite element simulation, the rolling head was defined as rigid body without deformation and strain. TC4 titanium alloy was set as the sample material, and cemented carbide was selected as the ultrasonic rolling head material. The specific material parameters are listed in table 3.

Table 3. Basic material parameters

Parameters	TC4	Rolling head
Density/kg.mm^{-3}	4.50E-6	1.46E-7
Elastic modulus/Gpa	110	510
Poisson's ratio	0.34	0.3
Yield strength/Gpa	1.098	-
Tensile strength/Gpa	1.092	-

In the process of ultrasonic rolling, the surface of TC4 material will deform strongly. Therefore, Johnson-Cook model is introduced to describe the mechanical properties. The equation is given in the formula(1)

$$\delta = (A + B\varepsilon^n)(1 + C\ln\dot{\varepsilon}^*)(1 - T^{*m}) \tag{1}$$

Where, δ is the yield limit; A is the yield stress; B is the strain power index coefficient; ε is the equivalent plastic strain; n is the strain hardening index; C is the strain rate sensitive coefficient; $\dot{\varepsilon}^*$ is the strain influence factor; $\dot{\varepsilon}_0$ is the reference strain rate; T^* is the temperature influence factor; m is the temperature sensitivity coefficient. Input the above parameters, melting point and reference temperature of the material into ABAQUS software. The specific parameter values are listed in table 4.

Table 4. Constitutive model parameters of TC4 material

A/MPa	B/MPa	n	m	melting point (K)	reference temperature (K)	C	$\dot{\varepsilon}_0$
1098	1092	0.93	1.1	1878	298	0.014	1

In the process of ultrasonic rolling, the contact condition changes with the change of machining position. The explicit module of ABAQUS is usually used to solve such nonlinear problems. The assembly model and mesh generation are displayed in figure 3. The rolling area is densely gridded with mesh transition on both sides, which weighs the operation speed and calculation accuracy.

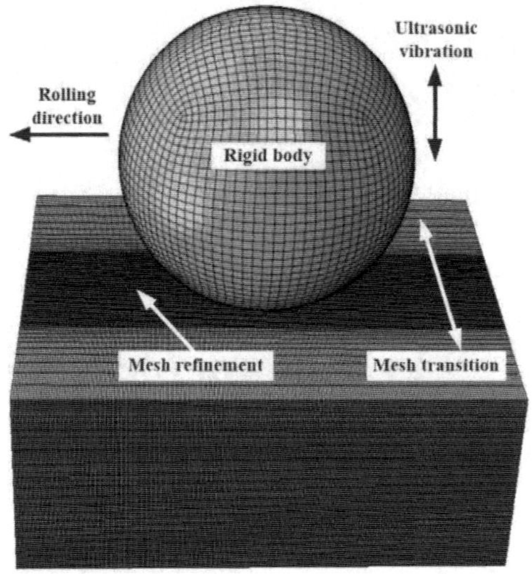

Figure 3. Assembly model and mesh generation.

3. Results and discussion

The surface residual stress nephogram obtained by finite element simulation is presented in figure 4. It can be seen from the figure 4 that a uniform residual stress layer with a certain depth is formed in the stable rolling zone. The residual stress values are extracted from the stable rolling zone along the depth direction in the figure. The comparison results of experimental values and simulation values are shown in figure 5.

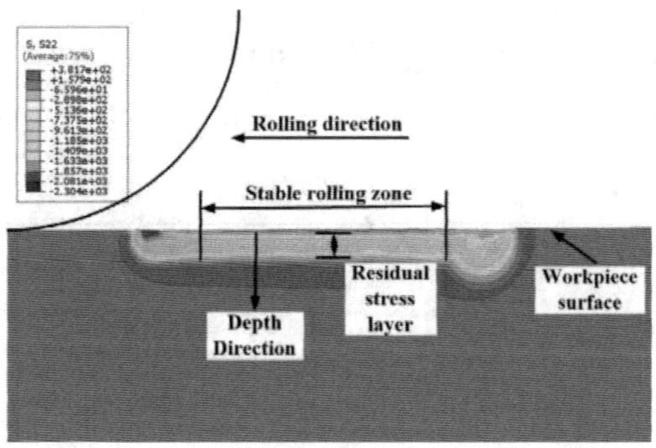

Figure 4. Surface residual stress nephogram.

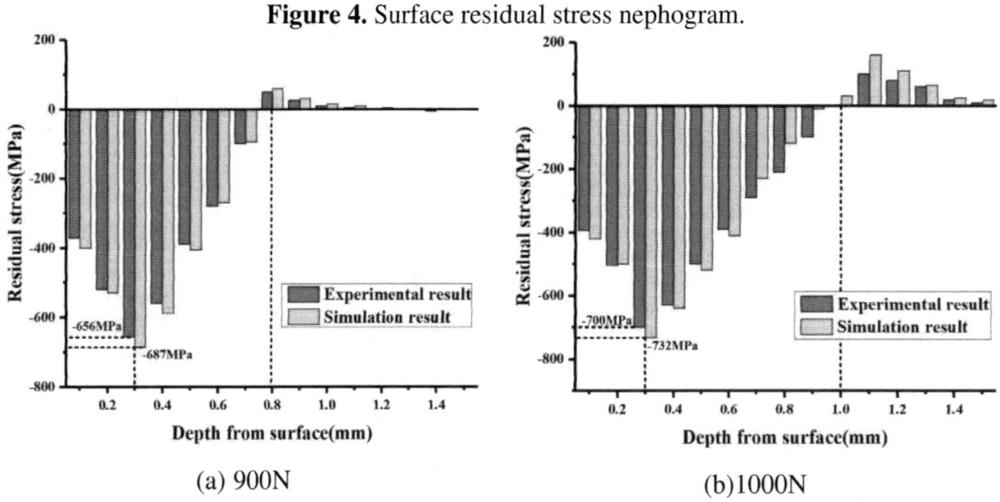

(a) 900N (b)1000N

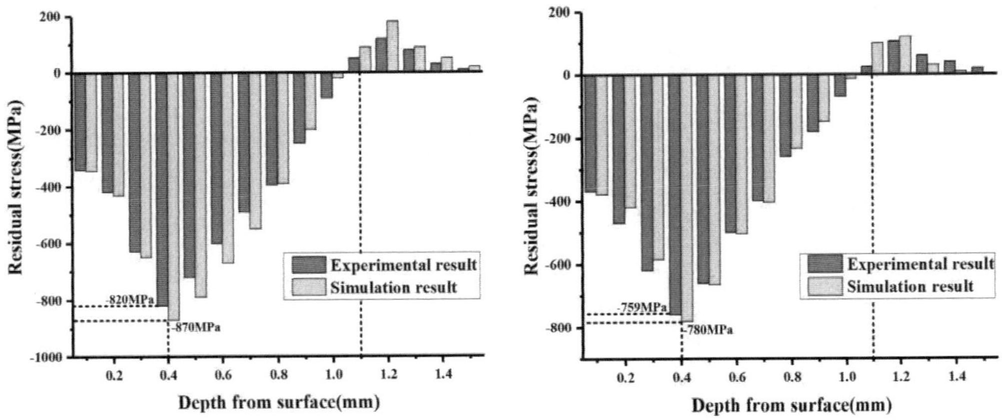

(b) 1100N (c) 1200N

Figure 5. Comparison results between experimental values and simulation values.

It can be seen from figure 5 that the residual stress reasults obtained by experiment and finite element simulation are basically the same. When the static pressure is 900N ~ 1000N, the position of the maximum residual compressive stress is about 0.3mm away from the surface. When the static pressure is 1100N ~ 1200N, it is about 0.4mm away from the surface.

When the static pressure is 900N, the residual compressive stress changes to the residual tensile stress at about 0.8mm from the surface. When the static pressure is 1000N, it changes to the residual tensile stress at about 1mm. When the static pressure is 1100N ~ 1200N, it changes to the residual tensile stress at about 1.1mm.

When the static pressure is 900N, 1000N, 1100N and 1200N, the maximum residual compressive stress values obtained by experiment are 656MPa, 699MPa, 820MPa and 759MPa respectively. The maximum residual compressive stresses obtained by finite element simulation are 686MPa, 732Mpa, 870MPa and 780MPa respectively. The errors are 4.7%, 4.6%, 6.1% and 2.8%, respectively, which improves the reliability of the results.

4. Summary

In this paper, the residual stress is measured by automatic gradient stress detection and analysis system. And then the ultrasonic rolling process of TC4 titanium alloy is simulated by ABAQUS software. The simulation and experimental curves of surface layer residual stress under static pressure of 900N, 1000N, 1100N and 1200N are obtained.

The experimental curve is roughly consistent with simulation curve. When the static pressure is 900N ~ 1000N, the position of the maximum residual stress is about 0.3mm from the surface. When the static pressure is 1100N ~ 1200N, it is located at the position about 0.4mm away from the surface.

When the static pressure is 1100N, the residual compressive stress reaches the maximum. The experimental value is 820MPa and the simulation value is 870MPa. The error between experimental value and simulation value is 6.1%. And the errors are 4.7%, 4.6% and 2.8%, respectively when the static pressure is 900N, 1000N and 1200N, which improves the reliability of the results.

5. References

[1] Yujing S, Jie S and Jianfeng L 2016 Finite Element Analysis on Prediction of Tool Wear in Milling Titanium J. Mech. Eng. JME05(2004)193

[2] Hexi J, Kexiang W, Jianming L, Jianyu Z and Wenjing P 2015 Research development of titanium alloy in aerospace industry Chin. J. Nonferrous. Met. CJNM02(2015)002

[3] Zongmin L, Bolin H and Yingxia Y 2017 Ultrasonic Impact Induced Surface Nano-crystallization and Influence on Surface Integrity *J. Mater. Sci. Eng.* JMSE02(2017)029

[4] Guanghui L, Youxi L and Jianguo C 2020 Research Progress on Ultrasonic Rolling Surface Strengthening of Metallic Material *Tool. Tech.* TE01(2020)001
[5] Yuan K and Sumi Y 2016 Simulation of residual stress and fatigue strength of welded joints under the effects of ultrasonic impact treatment (UIT) *Int. J. Fatigue.* IJF07(2016)018
[6] Jianxin Z and Shuxiang J 2017 Residual Stress Field in the Process of 2D Ultrasonic Rolling 7050 Aluminum Alloy *Surf. Tech.* ST12(2017)041
[7] Fengqin L and Bo Z 2019 Effect of Ultrasonic Processing Burnishing Pressure on Titanium Alloy Surface Properties *Surf. Tech.* ST10(2019)004
[8] Zhihua L, Tianzeng Z, Mengjian Y, Yinxia Z and Dong W 2020 Effect of Ultrasonic Rolling Parameters on Surface Properties of 18CrNiMo7-6 Gear Steel *Surf. Tech.* ST,03(2020)039

Acknowledgments

This work was supported by Major Projects for Transition to New Economic Engines from Old Ones in Shandong Province (SGG2001) and Major Innovation Projects in Shandong Province (No.2019JZZY010451)

IOP Conf. Series: Materials Science and Engineering 1063 (2021) 012002 doi:10.1088/1757-899X/1063/1/012002

Optimized Stress-Strain Ranges for Hyperelastic Constitutive Models Supporting the Simulation of Vertical Stiffness on Airless Tire

J Phromjan[1,2] and C Suvanjumrat[1,2]*

[1] Department of Mechanical Engineering, Faculty of Engineering, Mahidol University, Salaya, Nakhon Pathom, 73170, Thailand
[2] Laboratory of Computer Mechanics for Design (LCMD), Department of Mechanical Engineering, Faculty of Engineering, Mahidol University, Salaya, Nakhon Pathom, 73170, Thailand

chakrit.suv@mahidol.ac.th

Abstract. The airless tires are developed to overcome a disadvantage of pneumatic tire which lost maneuverability caused of flattening and blasting. The finite element analysis (FEA) is an effective method for development of airless tires which have a complex construction. The material properties which describe mechanical behavior of each tire component are very important to yield accurate result. The high order constitutive model usually obtains the accuracy of FEA results. Unfortunately, it consumes the simulating time. The suitable stress-strain range of Mooney-Rivlin and Ogden model was selected in this research. The vertical stiffness of airless tire model was compared to the experiment. This study found that the accurate result and consuming less time of TWEEL airless tire model depended on an appropriate pair of material properties. There were 100% of tread and 30% of spoke stress-strain characteristic based on the Mooney-Rivlin model. Consequently, the study and design of airless tire by finite element method will be carried out efficiently in the further works.

1. Introduction

The TWEEL airless tire 12N16.5 SSL ALL TERRAIN which was developed by Michelin for using on skid steer loader applications was designed to get better function than pneumatic tire without requirement of inflation pressure. The design of supported structure can provide the various advantages. To reduce the analytical development processes, finite element method was widely used to study tire behavior. Namjoo and Golbakhshi [1] had been studied the carcass stiffness of pneumatic tire by finite element method for tire safety development. The Mooney-Rivlin model was used to describe the material properties of tire model. The comparison of tire deformation agreed well with experimental result. The Yeoh model was used to define strain energy of the rubber material property of pneumatic tire model [2]. The 3D finite element model was used to study the tire noise generating mechanisms. The rolling behavior of tire model was very well with the published experimental results [3]. The finite element method had been also performed to investigate the fatigue damage of off-road tire. The Ogden model combining with fatigue crack growth law was used to define tire material properties [4]. The solid tire which was developed for support heavy load also was studied by 3D finite element method. To develop solid tire model, Phromjan and Suvanjumrat [5] had been varied hyperelastic models for analysis solid tire model deformation. They found that the Ogden model was the most agreement with the experiment. Moreover, the load capability of solid tire was studied by finite element method [6]. The airless tire was studied the mechanical behavior. Jin et.al, [7] had been performed the 3D finite element model of non-pneumatic tires (NPT) to investigate the static and dynamic behavior of various honeycomb spoke structures. The Ogden hyperelastic model and Prony series for polyurethane were used to define NPT material model. In the comparison of deformation, it is confirmed that the simulation results are reliable. Therefore, the NPT model can be used to study the stress distribution and rolling resistance. Rugsaj and Suvanjumrat [8] had been cut the TWEEL tire into material testing specimen according to ASTM standard. The 100% of stress-strain curve from material testing was used to fit with various hyperelastic model. The Ogden model was found to be the most suitable model for defining the material property of TWEEL airless tire. The thickness of spoke was varied. To obtain the same vertical stiffness with same size pneumatic tire, the spoke thickness of TWEEL tire could be reduced. Next, the accurate finite element model of airless tire was used to study the various spoke structure [9]. In the same way, the TWEEL airless tire model was developed by modification of steel belt layer to reduce the complexity of finite element analysis [10]. However, the complex structure of TWEEL tire was consumed a lot of computational resources and time. To optimize the hyperelastic model, this research will be varied the fitting percentage of stress-strain curve to reduce the simulating time and increase the accuracy of simulation results.

2. Experimental

2.1. Material testing

The airless tire, TWEEL 12N16.5 SSL ALL TERRAIN which is developed by Michelin is employed to studied in this research (Figure 1). It composed of rubber tread, shear band, belt layers and spoke. To obtain the material property of each component, the TWEEL airless tire was cut by the water jet cutting technique. The tread component was cut into cylindrical specimen accorded to the compression test standard, ASTM D575. The compression test on cylindrical specimen just performed to 25% of its height with compression speed of 12 mm/min. The spoke was cut into dumbbell specimen accorded to the tensile test standard, ASTM D412. The dumbbell specimen was pulled until it was broken with tensile speed of 500 mm/min. The Universal testing machine, Instron 5969, was carried out to investigate the material property of TWEEL airless tire components.

Figure 1. The airless tire TWEEL 12N16.5 SSL ALL TERRAIN.

2.2. Vertical stiffness testing of airless tire

The tire testing machine (Ektron PL-2003) is employed to perform the compression test on the TWEEL airless tire as shown in Figure 2. The vertical deformation and compressive force were recorded while the measurement table of testing machine was moved to press airless tire which was mounted on the mounting arm of the tire testing machine. The testing result indicated that the vertical stiffness of TWEEL airless tire by the compressive force of 14 kN was 932.88 N/mm.

Figure 2. The airless tire vertical stiffness testing by tire testing machine (Ektron PL-2003).

3. Finite element method

The hybrid formulation is effective for the analysis of rubberlike material in large strain [11]. The strain energy density per unit volume, ${}_0^t\bar{U}$, is defined in equation (1).

$$d\,{}_0^t\bar{U} = {}_0^t\bar{S}d\,{}_0^t\varepsilon_{ij} \tag{1}$$

where $d\,{}_0^t\bar{U}$ and ${}_0^t\bar{S}$ are incremental potential energy and Piola-Kirchhoff stress which computed only from the displace field. The integral form of total Lagrangian formulation is shown in equation (2).

$$\int_{V_0} {}_0^tS_{ij}\delta\,{}_0\varepsilon_{ij}dV_0 = R \tag{2}$$

The general form of principle of virtual work which written in the total Lagrangian formulation is given in equation (3).

$$\delta\left(\int_{V_0} {}_0^tUdV_0\right) = R \tag{3}$$

where ${}_0^tU$ is the incremental potential which can be modified to include the effect of interpolated pressure by adding to the term of displacement based on the total element pressure.

3.1. Hyperelastic model

The material property of rubberlike material is described by hyperelastic constitutive model which was explained by the strain energy function. The well-known hyperelastic model comprise of Neo-Hookean, Mooney-Rivlin, Ogden, Yeoh and Arruda-Boyce model. The Mooney-Rivlin model and Ogden model are most widely used to describe rubberlike material properties. They are written by equation 4 and 5, respectively [5].

$$U = C_{10}(\bar{I}_1 - 3) + C_{01}(\bar{I}_2 - 3) \tag{4}$$

$$U = \sum_{i=1}^{n} \frac{\mu_i}{\alpha_i}(\bar{\lambda}_1^{\alpha_i} + \bar{\lambda}_2^{\alpha_i} + \bar{\lambda}_3^{\alpha_i} - 3) + 4.5K(J^{1/3} - 1)^2 \tag{5}$$

where $\bar{\lambda}_i = J^{1/3}\lambda_i$, $J = \lambda_1\lambda_2\lambda_3$

C_{ij} is material parameter, $\bar{I}_1$ and $\bar{I}_2$ are the first and second invariants of the deviatoric strain.

λ_i is the deviatoric principle stretches, J is the Jacobean determinant, K is the initial bulk modulus and μ_i, α_i are material parameter.

3.2. Airless tire finite element model

The finite element model of airless tire was created according to TWEEL airless tire. It had 4 components which were tread, shear band, belt layer and spoke. The thick shell element was used to model spoke and belt layer of airless tire model while the hexagonal element was used to model tread and shear band. Figure 3(a) shows the finite element model of TWEEL airless tire. The airless tire model is combined with a rigid flat plate model to analyze the compression test by moving the flat plate model to press the fixed airless tire model as shown in Figure 3(b). The material property of tread and spoke were set following the constants of each constitutive model range to simulate the stiffness testing of airless tire. The vertical deformation and compressive force were recorded to compare with the experimental results of airless tire stiffness testing.

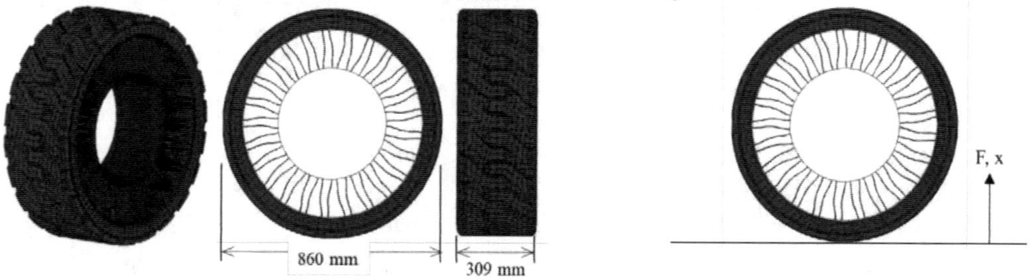

(a) (b)

Figure 3. The TWEEL airless tire (a) finite element model and (b) boundary condition of stiffness testing.

4. Results and discussions

The stress-strain relation of compression test and tensile test are presented in Figure 4(a) and 4(b), respectively. The stress-strain curves by the material testing of tread and spoke were divided into three ranges of stress-strain curve. The Mooney-Rivlin and Ogden constitutive model were fit to every stress-strain curve section of each airless tire component. Table 1 presents the constants of Mooney-Rivlin model by linear regression method of MSC.Marc software. The constants of Ogden model are presented in Table 2.

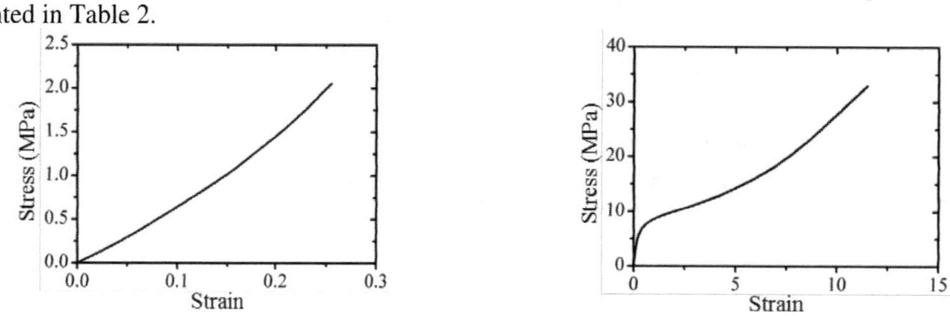

(a) (b)

Figure 4. The stress-strain relation of tread and spoke by: (a) compression test and (b) tensile test.

Table 1. The constant of Mooney-Rivlin constitutive model.

Curve section	Mooney-Rivlin's constant		R^2
	C10	C01	
	Tread		
30% of curve	0.99434	0.01565	0.9983
60% of curve	1.08977	0	0.9967
100% of curve	1.19055	0	0.9900
	Spoke		
30% of curve	0.13031	4.95221	0.9793
60% of curve	0.53117	3.92669	0.9579
100% of curve	0.8388	2.88528	0.9721

Table 2. The constant of Ogden constitutive model.

Curve section	Ogden's constant					R^2
	μ_1	μ_2	α_1	α_2	K	
			Tread			
30% of curve	0.00732	15.8447	38.8412	0.22534	9636.99	0.9999
60% of curve	0.45942	5.48E-08	8.25776	3.51768	9484.52	0.9998
100% of curve	0.52047	30.4738	7.43197	1.44E-09	9670.24	0.9998
			Spoke			
30% of curve	0.00039	566.929	7.08047	0.03107	44037.4	0.9842
60% of curve	590.136	0.08884	0.02825	3.38497	42433	0.9626
100% of curve	5760.18	0.20123	0.00277	2.96849	41425.4	0.9989

The model fitting at the low percentage of stress-strain curve was more precise than the high percentage because of the stress-strain curve trended to be linear in a little deformation for the compression test of tread. The stress-strain curve of spoke was non-linear then the precision is varied. The TWEEL airless tire model which was governed by Mooney-Rivlin and Ogden constitutive model was compressed by the vertical load of 14,000 N. Table 3 presents comparison results of the vertical stiffness between finite element method and experiment.

Table 3. The comparison result of vertical stiffness of airless tire model with experiment.

Case No.	Material	Mooney-Rivlin			Ogden			Simulation time (min.)	%error
		30%	60%	100%	30%	60%	100%		
1	Tread	•						32.12	15.88
	Spoke	•							
2	Tread	•						34.92	19.72
	Spoke		•						
3	Tread	•						34.15	24.82
	Spoke			•					
4	Tread		•					31.04	15.27
	Spoke	•							
5	Tread		•					30.58	19.18
	Spoke		•						
6	Tread		•					28.46	24.37
	Spoke			•					
7	Tread			•				31.60	14.64
	Spoke	•							
8	Tread			•				34.78	18.61
	Spoke		•						
9	Tread			•				34.26	23.85
	Spoke			•					
10	Tread				•			29.75	19.26
	Spoke				•				
11	Tread				•			31.10	20.35
	Spoke					•			
12	Tread				•			32.00	21.07
	Spoke						•		
13	Tread					•		33.18	20.41
	Spoke				•				
14	Tread					•		33.68	21.45
	Spoke					•			
15	Tread					•		34.65	22.17
	Spoke						•		
16	Tread						•	34.07	20.33
	Spoke				•				
17	Tread						•	31.86	21.39
	Spoke					•			
18	Tread						•	32.39	22.09
	Spoke						•		

The lowest error of vertical stiffness simulation occurred at the case number 7. The simulation time of this case was 31.60 min was not much when compared to the others and was less than simulation time of the case number 9 and 18 which were the 100% of fitting. Total results of 18 simulating cases were found that the accurate result of TWEEL airless tire was depended on the appropriate pair of material properties. Figure 5 shows the airless tire deformation. The color contour presented the value of displacement of TWEEL airless tire model. The yellow color was the maximum value of displacement while blue was minimum. The simulation result presented the most deformation of TWEEL airless tire which occurred at the tread and spoke while they were compressed. The spoke was bended distinctly when it rolled to be normal with the horizontal floor.

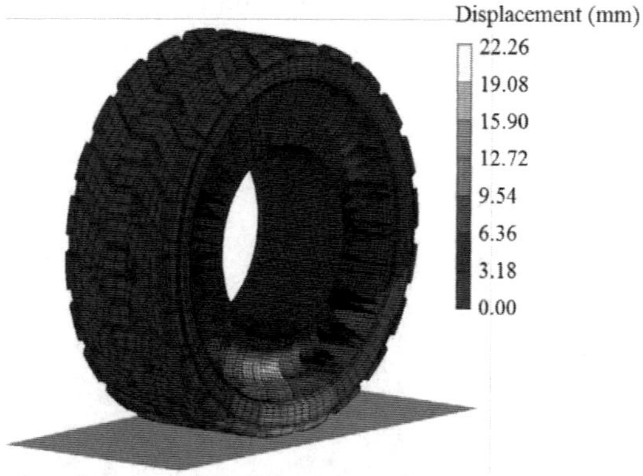

Figure 5. The total displacement of TWEEL airless tire model by simulation of case number 7.

5. Conclusion

The TWEEL airless tire components which were composed of tread, shear band, belt later and spoke had been carried out to test the material property. The stress-strain relation of these components was fit with Mooney-Rivlin and Ogden hyperelastic model. The constitutive models of each stress-strain range were implemented to the stiffness testing of TWEEL tire model. The accurate result of TWEEL airless tire was depended on an appropriate pair of material properties. It was found that the pair of 100% of tread and 30% of spoke stress-strain curve fitting based on the Mooney-Rivlin model provided the lowest error of 14.64% while the simulation time was 31.60 mins. The shape of spoke was designed to be deformed when the spoke rolled to be normal with the floor. The method of constitutive model selective to develop airless tire model in this research can be used to study and design the airless tire by finite element analysis under the appropriate accuracy and time in future work.

6. References

[1] Namjoo M and Golbakhshi H 2015 *J. Cent. South Univ.* **22** 189
[2] Palanivelu S, Narasimha R and Ramarathnam K K 2015 *Mech. Syst. Signal Process* **64-65** 385
[3] Kindt P, Sas P and Desmet W 2009 *Opt. LasersEng.* **47** 443
[4] Nyaaba W, Frimpong S and Anani A 2019 *Int. J. of Fatigue* **119** 247
[5] Phromjan J and Suvanjumrat 2018 *Engineering journal* **22(2)** 141
[6] Phromjan J and Suvanjumrat C 2018 *Key Eng. Mater.* **777** 416
[7] Jin X, Hou C, Fan X, Sun Y, Lv J and Lu C 2018 *Compos. Struct.* **187** 27
[8] Rugsaj R and Suvanjumrat C 2019 *Int. J. Automot. Technol.* **20(4)** 801
[9] Rugsaj R and Suvanjumrat C 2020 *Mech. Based Des. Struc.* DOI:10.1080/15397734.2020.1777875
[10] Phromjan J and Suvanjumrat C 2020 *IOP Conf. Ser-Mat Sci.* **773(1)** 012047
[11] Bathe K J 1997 *Finite Element Procedures* (London: Prentice Hall)

IOP Conf. Series: Materials Science and Engineering 1063 (2021) 012003 doi:10.1088/1757-899X/1063/1/012003

Finite Element Analysis of Notch Sensitivity of PEEK under Monotonic Tension

M A A Khan*, M Asad, F Djavanroodi, J F Nayfeh, and T Waqar

Mechanical Engineering Department, College of Engineering, Prince Mohammad Bin Fahd University, Al Khobar, Kingdom of Saudi Arabia

*E-mail: mkhan6@pmu.edu.sa

Abstract. Polyetheretherketone (PEEK) is of great interest to the biomedical industry especially for trauma, orthopedic and spinal implants. It is biocompatible and offers numerous advantages over other metallic biomedical implants. A good understanding of behavior of PEEK under different loading conditions is therefore necessary for its more efficient use. This paper investigates the notch sensitivity of PEEK in monotonic tension using the finite element modeling in ABAQUS. An unnotched specimen and three notched specimens (moderate, deep and razor) are modeled and the maximum true stress in each case is obtained. The stress distribution in front of the crack tip is also found by introducing a crack in the model for each specimen. It is observed that PEEK exhibits a ductile behavior in the absence of a stress concentration. The deformation mechanism is found to be dramatically changed from ductile to brittle in the presence of notches. Hence, the locations and severity of stress risers modeled as notches in PEEK should be taken into consideration prior to its use in biomedical implants.

1. Introduction

Polyetheretherketone (PEEK) is a high performance thermoplastic polymer [1] which has emerged as one of the most widely used biomaterials over the last two decades. Owing to its biocompatible nature, it has remained a material of interest to the medical community in trauma, orthopedic, dental, spinal and cranial implants [2–4]. It is a semi crystalline polymer with about 30-35 % crystallinity and glass transition temperature and crystalline melt temperature of 143 °C and 343 °C, respectively [2]. Besides PEEK'S high strength and good load bearing capabilities, it also offers other advantages such as being X-rays transparent, creates no artifacts in CT images, and shows excellent biocompatibility [5]. Moreover, PEEK confers chemical and high temperature stability, compatibility with many reinforcing agents (such as glass and carbon fibers), and exhibits higher strength (on a per mass basis) than many metals.

Unfilled or "neat" PEEK is generally available in powder, pellets, extruded rods, and films as shown in Figure 1. Pellets and powders can be converted to implants using any polymer processing technique such as injection molding. Alternatively, these implants can be machined from extruded rods and compression molded sheets [1]. When combined with any reinforcing material such as hydroxyapatite (HA) and/or carbon fibers, they become PEEK composites and as such, their properties can be tailored by changing the composite formulation. While HA/PEEK composites are considered to be appropriate for bone replacement due to their micro and macro enhanced mechanical properties as well as their biocompatibility, it is important to note that their ductile and tensile strength properties are often compromised by the increased content of the HA particles and due to the poor interfacial adhesive bond [6]. Similarly, carbon fiber reinforced (CFR) PEEK composites are currently employed in implants involving spine fusion and joint replacement orthopedic procedures. Their directional micro and macro structural properties as a composite material are dependent on the fibers characteristics including their material type and size, length and orientation within the host matrix material.

Figure 1. Powder, pellets, rods, and film of neat PEEK [1].

Since PEEK is being used substantially for human implants, it is of utmost importance to be familiar with the failure behavior of this material under different loading conditions. Due to the intricate complexities of its functions as a structural prosthesis in a human host body, its long term performance and effects continue to be of paramount importance to the materials, mechanics and medical research communities. This study investigates by introducing predetermined geometric defects in the form of notches that simulate cracks in PEEK structures subjected to monotonic tension using the finite element modelling in ABAQUS. Background and experimental details are presented in this paper. Details of the ABAQUS finite element model simulations and results are discussed.

2. Background

The effects of stress concentrations in implants made up of PEEK biomaterials are important and must be carefully analyzed in a multitude of its applications. Sobieraj et al. [7] studied these effects by examining both stress-strain and fracture behavior of PEEK structures under uniaxial loading in circumferentially grooved round bar specimens with different elastic stress concentration factors. Specimens were prepared with neat PEEK formulation. In total four specimens were tested out of which one specimen was un-notched, and the other three had moderate, deep and razor notches as shown in Figure 2 along with their radii and elastic stress concentration factors (K_t). In order to replicate physiological conditions, the specimens were soaked in a 37 °C phosphate-buffered saline bath for 8 weeks.

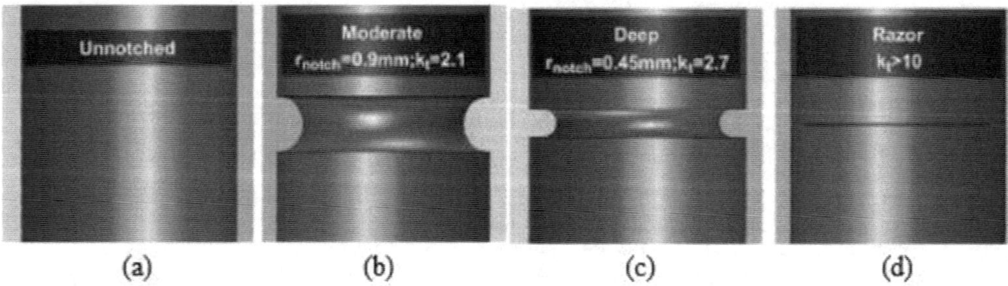

Figure 2. Specimen geometries used in experiments (a) Unnotched (b) Moderate notch (c) Deep Notch (d) Razor Notch.

Two displacement rates i.e. 6 mm/min and 30 mm/min have been used for testing. Load-displacement data was recorded in a conventional manner using lab view, however, strain data was recorded using a non-intrusive video based method. True axial stress-strain curves were presented for un-notched and notched specimens, also von Mises stress-strain curves were presented for the un-notched specimen. The un-notched specimen experienced considerable necking prior to fracture in contrast to the U-shaped notched specimens which were found to be brittle under the same loading conditions as shown in Figure 3. Stress-strain curves (axial true and Von-Mises) confirmed significant strain hardening in the necking region as shown in Figure 4. In addition to this axial stress-strain curves presented in Figure 5, the figure shows deviation from linear behavior before failure confirming some plastic behavior in the specimen.

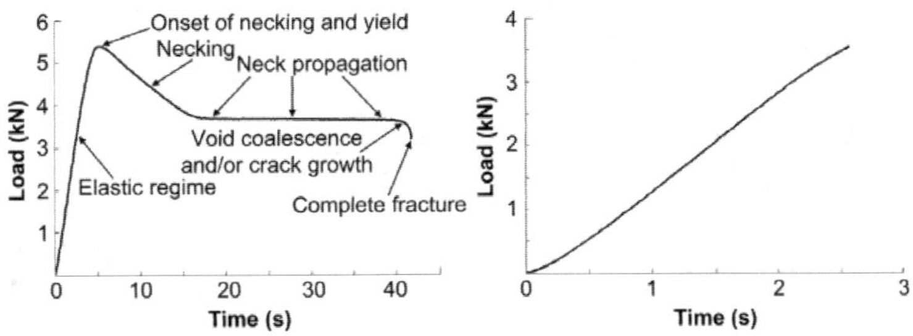

Figure 3. Load-time curves for an un-notched specimen (left, 6mm/min) and a deep notched specimen (right, 6 mm/min).

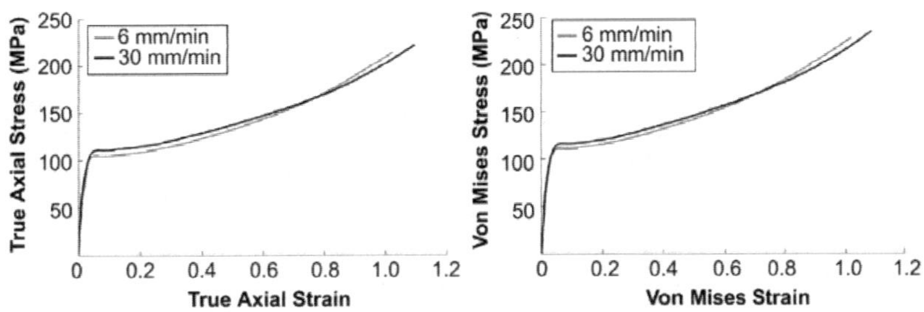

Figure 4. Axial true stress–strain (left) and von Mises stress–strain (right) curves for the un-notched geometry.

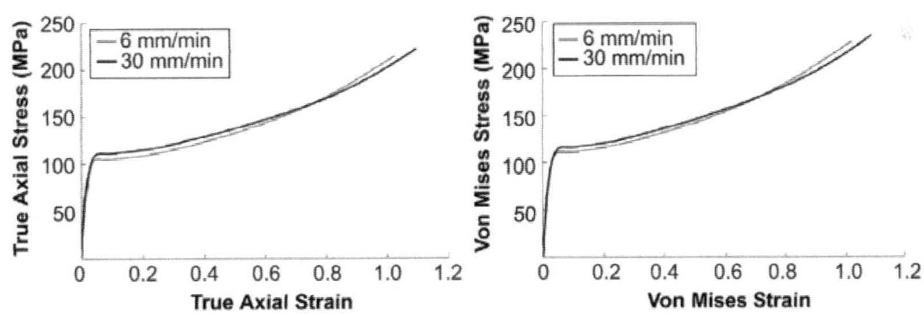

Figure 5. Axial true stress–strain for the moderate and the deep notched specimens at both displacement rates [7].

3. Methodology

In this study, four type of specimen geometries presented in Figure 2 are modeled in ABAQUS. The details of modeling and simulation are discussed in the forthcoming sub-sections.

3.1. Modelling of Notch in Specimens

The modelling of specimen is done using the exact dimensions as provided in Figure 2. Three dimensional models of all specimens' type (i.e., unnotched, moderate notch, deep notch, and razor notch) are developed in ABAQUS as shown in Figure 6. The PEEK material is modelled as an elastic-plastic material with the properties shown in Table 1. The specimens are discretized using a *Tet mesh* with a minimum of 0.4 mm node spacing. This node spacing was used because the changes in the simulation results are insignificant upon further mesh refinement.

IOP Conf. Series: Materials Science and Engineering 1063 (2021) 012003 doi:10.1088/1757-899X/1063/1/012003

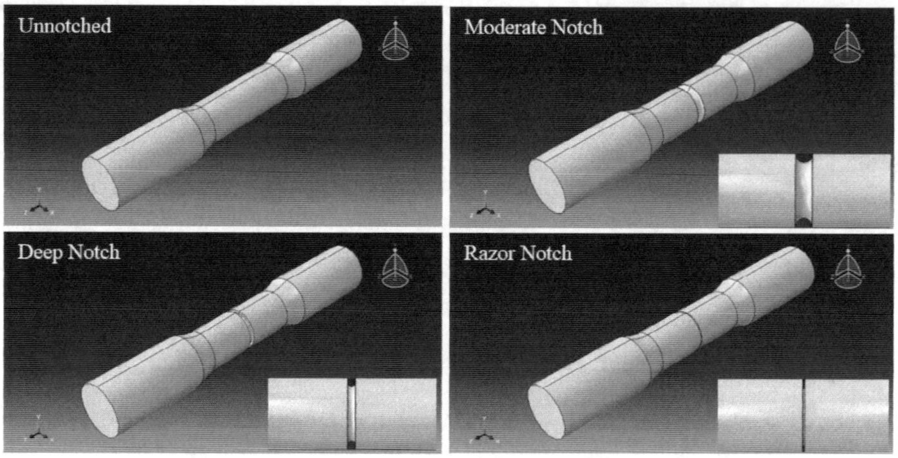

Figure 6. Modelling of PEEK specimens in ABAQUS.

Table 1. Mechanical Properties of PEEK [7], [8].

Material Property	Unit	Value
Density	Kg/m^3	1300
Elastic Modulus	MPa	3500
Poisson Ratio	-	0.36

Since the specimens are loaded in monotonic tension in each case, the specimens are fixed at one end and the other end of the specimen is loaded under displacement control. As discussed earlier the experiments were performed at two displacement rates i.e. 6 mm/min and 30 mm/min [7]. However, for simplicity, only one displacement rate 6 mm/min was selected to simulate the test. The load vs. time curves for the unnotched and deep notched specimens, depicted in Figure 3, were used to calculate the displacement at failure. No information about displacement at failure for the moderate notched and razor notched specimens were provided. Nevertheless, it was reported that in the moderate notched specimen, a large area experienced plastic deformation when compared with the deep notched specimen. Table 2 summarizes the boundary conditions for displacement in all four specimens at 6 mm/min displacement rate.

Table 2 Boundary conditions for different specimens

Specimen Type	Displacement (mm)
Unnotched	4.2
Moderate Notch	0.5
Deep Notch	0.25
Razor Notch	0.13

3.2. Modelling of Crack in Specimens

Notches are the areas of stress concentrations in a material and therefore it is highly probable that structural cracks may initiate and propagate form that region. For the purpose of this study, a 1 mm long crack is modelled on a single side of each of the four different specimens. For the unnotched specimen, the crack is modelled at exactly the centre of the specimen, however, for the notched specimens, the crack is modelled such that it starts at the notch radius which corresponds to the surface of the inner diameter of the specimen. Figure 7 presents the geometrically modelled 2-D cracks of the four specimens. The cracks in all four specimens are modelled using the same exact material properties and boundary conditions as discussed in Section 3.1.

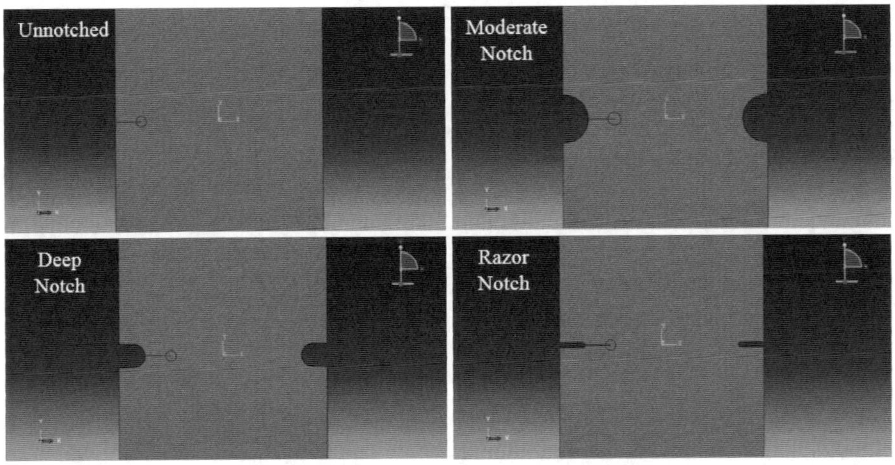

Figure 7. Modeling of crack in PEEK specimens.

4. Results and Discussions

Figure 8 presents the simulated results for all four specimens. It can be observed that the unnotched specimen experienced significant necking. Maximum true stress is observed in the necked region of this unnotched specimen, and also at the notch radii of the moderate, deep and razor notched specimens. Table 3 summarizes the experimental and simulated maximum true stress values calculated in all four specimens. Figure 9 presents the trend of maximum stress observed in the four specimens for both experimental and simulated conditions. In both cases, the maximum and minimum true stresses are observed in the unnotched and razor notched specimens, respectively. Although the difference between the experimental and simulated values cannot be ignored, it is important to note that the material behaved similarly in both conditions. These discrepancies in the quantitative results can be attributed to possible actual differences in the material modeling parameters and minor mismatches in the boundary conditions between the ABAQUS simulation model and the physical experimental setup as well as the ABAQUS meshing geometries and parameters. The failures in the computational model were nevertheless replicating and confirming qualitatively the experimental conditions and results.

Table 3. Experimental and simulated values of maximum true stress in different PEEK specimens.

Specimen Type	Maximum Stress (MPa)	
	Experimental	Simulated
Unnotched	211	143
Moderate Notch	132	94
Deep Notch	127	89
Razor Notch	120	77

Both experimental and simulated results provided insights into the mechanical behaviour of PEEK specimens with and without stress concentrations. The research findings document that PEEK undergoes large plastic deformation and is more ductile under unnotched condition. Figure 10 compares the necking behaviour of PEEK in both experimental and simulated conditions which are found to be in good agreement. With the introduced notch geometries as stress risers in the specimens, PEEK starts to exhibit brittle behaviour with very small plastic deformation as already depicted in

Figure 5 [7]. As reported by [7], ~2% reduction in the diameter of the specimen with notches was observed.

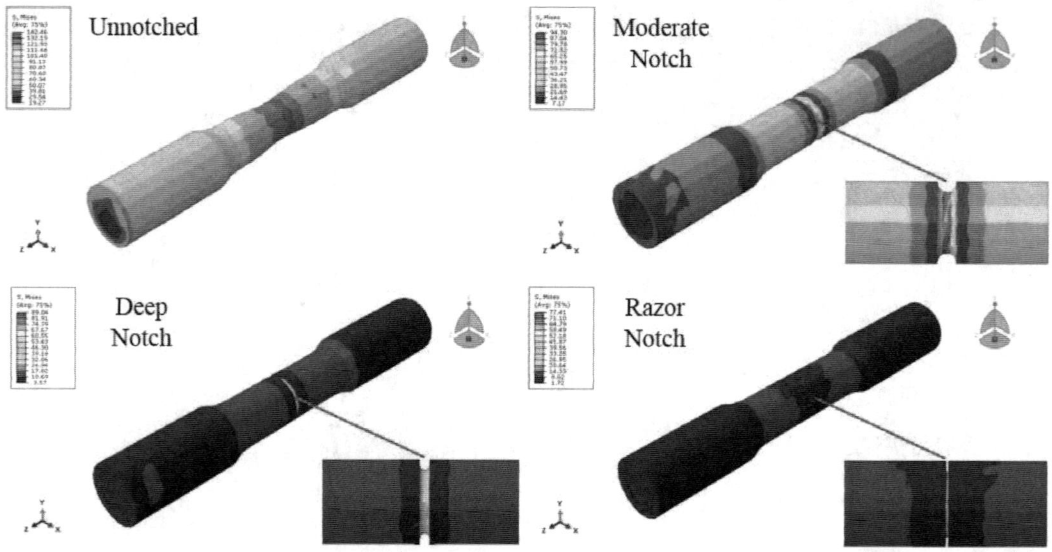

Figure 8. Simulation results of unnotched and notched PEEK specimens under tension.

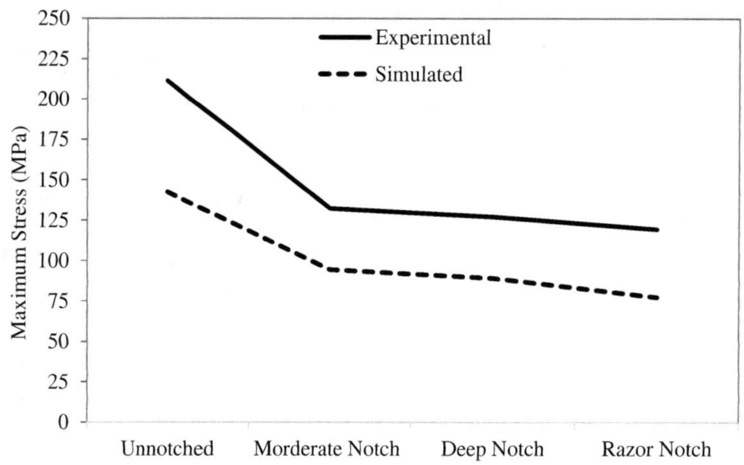

Figure 9. Experimental and simulated maximum true stress in different PEEK specimens.

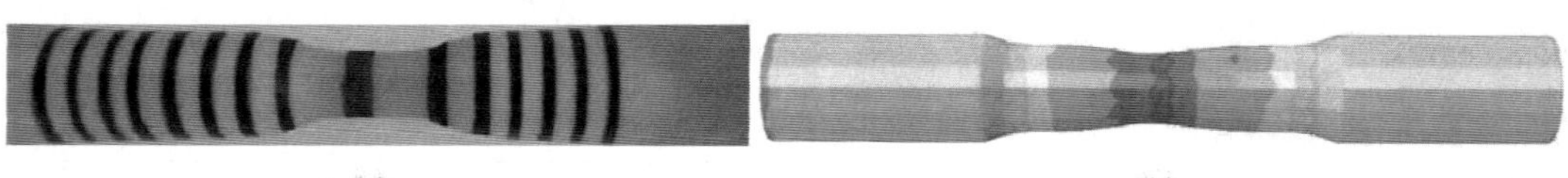

Figure 10. (a) Experimental and (b) Simulated necking of unnotched PEEK specimen.

The propagating neck and thus the ductile behaviour of unnotched PEEK specimen can be explained by the increased chain alignment at the surface of the neck compared to the interior, a typical situation for polymers that undergo necking while deforming axially [9]. Sobieraj et al. [7] also explained the whitening observed during deformation as a precursor to void coalescence. Moreover, the dip observed in the load-time curve as shown in Figure 3 is representative of initiation of fracture due to void coalescence and crack growth for an internally initiated failure or crack growth from an external initiation site [7].

As discussed earlier, notching in the specimens dramatically change the behaviour from a highly ductile to a brittle material. Sobieraj et al. [7] explained this behaviour with respect to cavitation. Cavitation is associated with the hydrostatic stress present in a polymer and occurs when this reaches a critical cavitation stress. Von Mises stress governs the yielding in a specimen. Semicrystalline polymers under tension experience a competition between cavitation during elastic deformation and activation of crystal plasticity. Elastic cavitation occurs in semicrystalline polymers with crystals of higher resistance to the onset of plastic behaviour, while plastic deformation occurs in crystals of lower plastic resistance. With unnotched specimen under consideration, yielding and then plastic flow predominated, however, notching suppressed the yielding due to which elastic behaviour and cavitation became a dominant phenomenon.

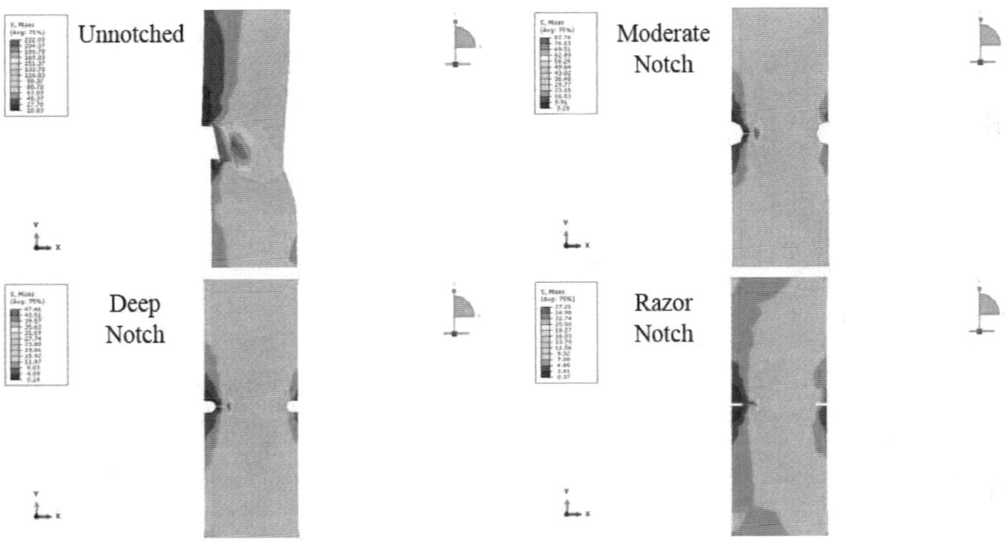

Figure 11. Simulation results of unnotched and notched PEEK specimens with initial crack of 1 mm.

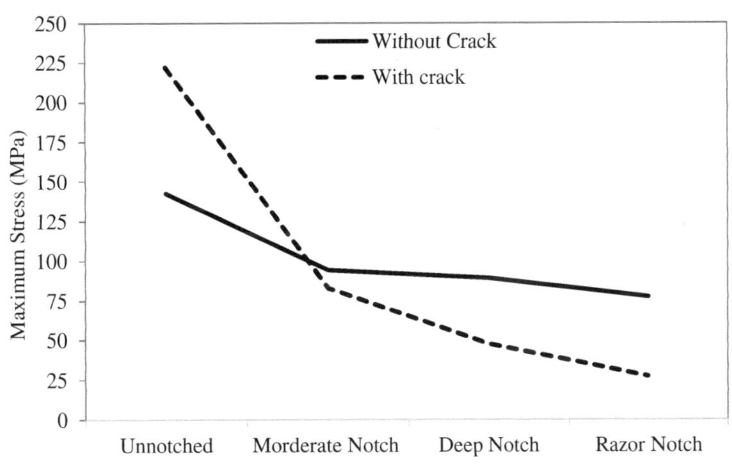

Figure 12. Experimental and simulated maximum true stress in unnotched and notched PEEK specimens with crack.

IOP Conf. Series: Materials Science and Engineering 1063 (2021) 012003 doi:10.1088/1757-899X/1063/1/012003

Figure 11 presents the stress distribution within all test specimens in the presence of a 1 mm crack which is modelled in the centre of the unnotched specimen and on the point of maximum stress concentration in the notched specimens. The unnotched specimen is found to be severely distorted in the presence of the crack and very high values of stress are obtained within the specimen. This increased stress in the unnotched specimen needs to be studied further. For each of the notched specimens, the value of the maximum true stress as shown in Figure 12, is found to be lower than in the cases simulated when there was no crack in the specimen. This suggests a further decline in the specimen ductility in the presence of the crack.

5. Conclusions

The notch sensitivity of PEEK is studied using finite element modelling in ABAQUS. Unnotched and moderate, deep and razor notched specimens are modeled and compared for the maximum true stress already calculated using experimental data. The experimental and simulated maximum true stress followed the same trend in unnotched and notched specimens. However, discrepancies are observed in the magnitude of maximum true stress in the experimental and simulated results. It was observed that notching caused a significant suppression in the overall ductility of the specimen and thus altered the deformation behavior. Next, a crack is modeled in each of the specimen and stress distribution around the crack tip is obtained. The findings of this study emphasizes on the careful consideration of the locations and severity of notch risers in biomedical implants utilizing PEEK.

The study can be extended further by improving the finite element modelling of PEEK in terms of material definition, meshing parameters and replicating the actual test conditions. Also, the propagation of an existing crack in PEEK can be studied using extended finite element method (XFEM). It is also important to note that biomedical implants are not only subjected to tensile loading, and thus, the behaviour of PEEK should also be studied under other loading conditions such as torsion, impact, and cyclic loadings etc.

References

[1] Kurtz S M 2012 Chapter 1 - An Overview of PEEK Biomaterials *PEEK Biomaterials Handbook* Plastics Design Library ed S M Kurtz (Oxford: William Andrew Publishing) pp 1–7

[2] Kurtz S M and Devine J N 2007 PEEK biomaterials in trauma, orthopedic, and spinal implants *Biomaterials* **28** 4845–69

[3] El Halabi F, Rodriguez J F, Rebolledo L, Hurtós E and Doblaré M 2011 Mechanical characterization and numerical simulation of polyether–ether–ketone (PEEK) cranial implants *Journal of the Mechanical Behavior of Biomedical Materials* **4** 1819–32

[4] Lovald S and Kurtz S M 2012 Chapter 15 - Applications of Polyetheretherketone in Trauma, Arthroscopy, and Cranial Defect Repair *PEEK Biomaterials Handbook* Plastics Design Library ed S M Kurtz (Oxford: William Andrew Publishing) pp 243–60

[5] Williams D F, McNamara A and Turner R M 1987 Potential of polyetheretherketone (PEEK) and carbon-fibre-reinforced PEEK in medical applications *J Mater Sci Lett* **6** 188–90

[6] Fan J P, Tsui C P and Tang C Y 2004 Modeling of the mechanical behavior of HA/PEEK biocomposite under quasi-static tensile load *Materials Science and Engineering: A* **382** 341–50

[7] Sobieraj M C, Kurtz S M and Rimnac C M 2009 Notch sensitivity of PEEK in monotonic tension *Biomaterials* **30** 6485–94

[8] Garcia-Gonzalez D, Rusinek A, Jankowiak T and Arias A 2015 Mechanical impact behavior of polyether–ether–ketone (PEEK) *Composite Structures* **124** 88–99

[9] Cao J 1992 Studies on the mechanism of the instability of polymers in tensile testing *J. Appl. Polym. Sci.* **45** 2169–79

Acknowledgments

The authors would like to acknowledge the support provided by Prince Mohammad Bin Fahd University in this research.

IOP Conf. Series: Materials Science and Engineering 1063 (2021) 012004 doi:10.1088/1757-899X/1063/1/012004

The Dynamic Characteristic Analysis of an A/B Biaxial Rotary Milling Head

Bin Tian[1], Guangshun Liang[2] and Yun Zhang[2,*]

[1]Neijiang Jinhong crankshaft Co., Ltd, Neijiang, Sichuan Province, China
[2]Beijing Key Lab of Precision/Ultra-precision Manufacturing Equipments and Control, Tsinghua University, Haidian District, Beijing, China

*E-mail: zhangyun@tsinghua.edu.cn

Abstract. The biaxial rotary CNC universal milling head is an important functional part of the five-axis CNC machine tool, and its dynamic performance directly affects the machining accuracy of the machine tool. In this paper, the modal analysis and thermal error analysis of an A/B biaxial rotary milling head were carried out. The first six natural frequencies and modal shapes were obtained by the finite element method, and the modal test was carried out by LMS Test.Lab. The transient temperature field of the milling head during machining was obtained by the finite element method, and the thermal deformation experiment was carried out. The experimental results verify the validity of the finite element model within an acceptable error range. This paper provides theoretical references for the structural optimization and thermal error compensation of the A/B biaxial rotary milling head.

1. Introduction

In the field of aerospace manufacturing, one of the most concerned functions of high-end CNC machine tools is five-axis machining. To achieve this function, biaxial rotary CNC universal milling heads are indispensable. In the process of the biaxial rotary milling head, in addition to the steady-state force, it is also affected by other factors such as alternating cutting force, cutting heat, thermal deformation, environmental temperature and so on. Sources of the processing error mainly include vibration and thermal deformation [1]. Vibration occurs when the milling head is subjected to alternating loads. When the load frequency is close to the natural frequency of the milling head, it is prone to resonance, resulting in deterioration of the performance of the machining system and control system. Various thermal loads during processing make the temperature rise of each part of the milling head asymmetric, result in varying degrees of thermal deformation of the structure and affect the processing accuracy. It is of great significance to improve the machining accuracy of the milling head by investigating its long-term working characteristics under the dynamic and thermal loads, that is, the dynamic characteristics of the milling head, so as to improve the performance of the five-axis machine tool and the level of the national manufacturing industry.

In recent years, many scholars have studied the dynamic characteristics of various milling heads by means of finite element simulation or field experiment, and obtained the corresponding parameters, which provides the basis for the structural optimization and thermal error compensation of milling heads. Bao (2012) introduced the model of the direct-drive A/C biaxial rotary milling head into ANSYS, and established the thermal error compensation system based on the thermal characteristics of the milling head. He also carried out the overall static analysis and modal analysis of the milling head, providing the basis for the improvement and optimization of the milling head structure [2].

Wang (2016) collected the natural frequencies, damping ratios and modal shapes of the first six modes of the woodworking biaxial rotary milling head by using the PULSE multi-analyser system, and verified the vibration resistance performance of the milling head [3].

The CNC machine tool with the A/B biaxial rotary milling head can complete the task of high-power and high torque cutting, and has the characteristics including unique tool path and simple equipment operation, which plays an important role in the manufacturing of large-scale integral aircraft structural parts with a long processing cycle. However, the research on the dynamic performance of the A/B biaxial rotary milling head is relatively less at present. This paper studies the dynamic characteristics of an A/B biaxial rotary milling head, including the modal analysis and thermal error analysis, in order to provide data support for the design and improvement of this type of milling head.

2. Modal analysis of the A/B biaxial rotary milling head

2.1. Modal finite element analysis

The schematic diagram and three-dimensional model of the milling head used in this research are shown in figure 1. The overall size is 1550mm × 830mm × 930mm. Its main structure includes axis A drive, axis B drive, main drive, spindle box, support frame and other parts. The axis A performs the swing of the spindle head around the axis X. And the axis B performs the swing of the spindle head around the axis Y. The servo motor drives the gear through the belt and worm gear, and finally drives the axis A and the axis B on the arc rack. The ratio of angular velocity of the axis A to the axis B is 1127.8

(a)

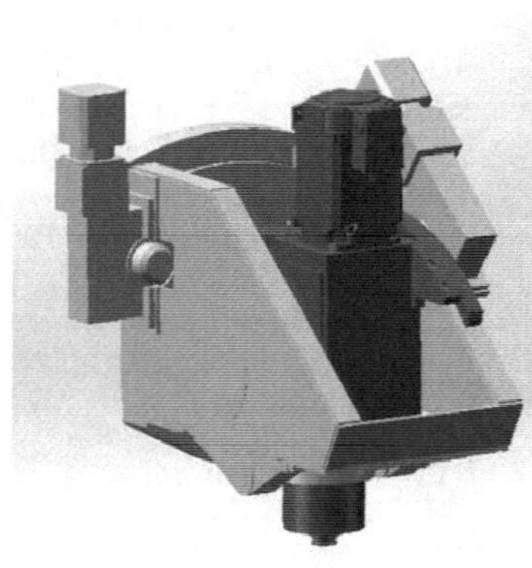
(b)

Figure 1. (a) Schematic diagram (b) 3D model.

In this study, the materials of the milling head are nodular cast iron and alloy structural steel, of which the spindle box is made of cast iron QT600-3, and the others are made of alloy structural steel. Among them, gears are made of alloy steel 12CrNi3A, the main shaft is made of alloy steel 42CrMo, bearings are made of bearing steel GCr5, and other parts are made of steel 45#.

The model was simplified in ANSYS Workbench, and the mesh was divided by automatic division method. After division, the number of generating units was 187330 and the number of nodes was 548754. Check the quality of the grid, and the average value was 0.78. It can be considered that the quality of the grid was satisfactory. Six degrees of freedom of the end face of the milling head were constrained, and the first six modes were extracted. The modal shapes are shown in figure 2, and the natural frequencies and characteristics of the modal shapes are shown in table 1.

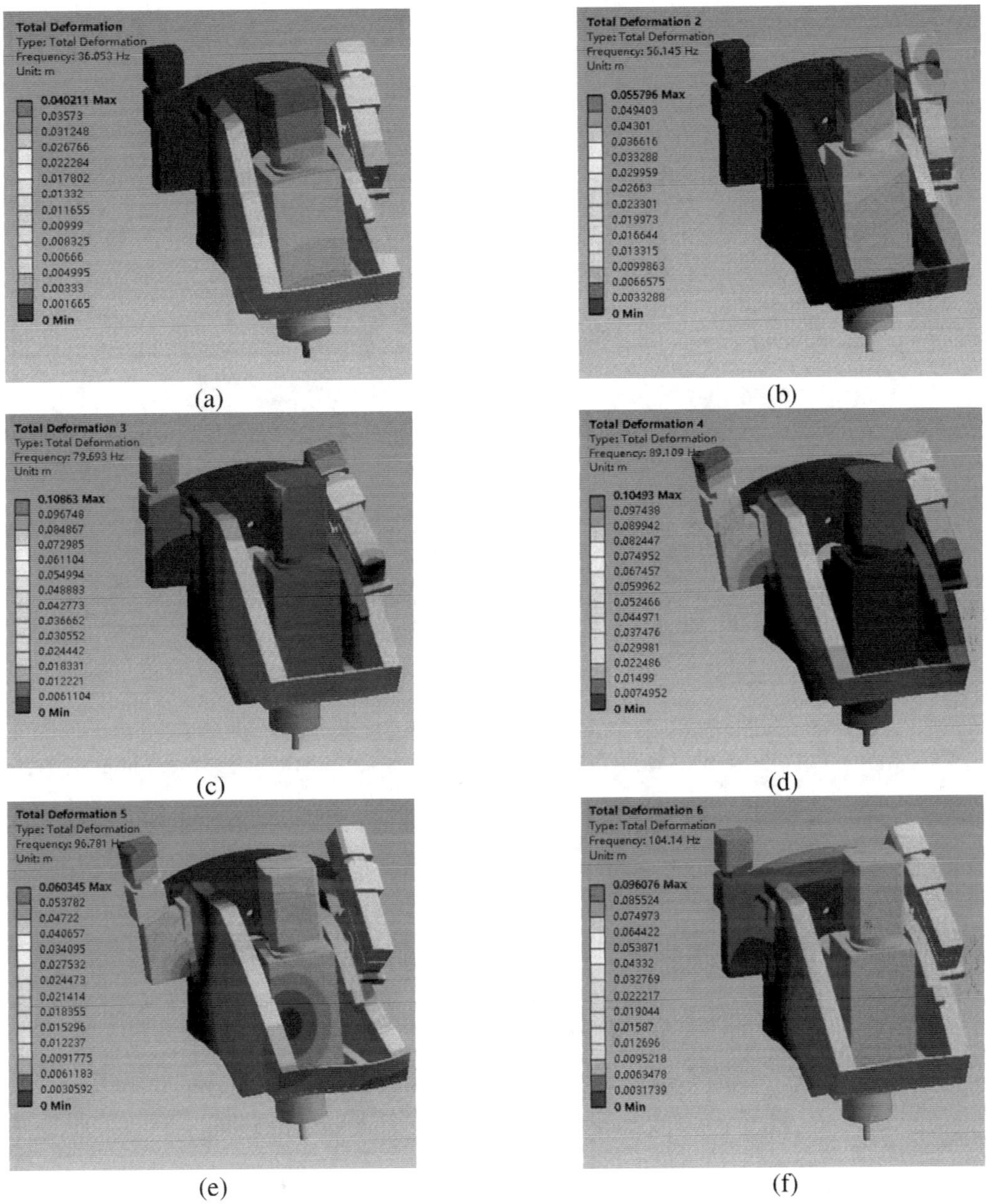

Figure 2. Modal shapes of the milling head (a) first order (b) second order (c) third order (d) fourth order (e) fifth order (f) sixth order.

Generally, the low-order modes have great influence on the structure. As is shown in table 1, the first two natural frequencies of the milling head are low. And both of them are within 60Hz, which causes the chatter of the milling head easily during processing. It can be seen from the modal shapes that the reason for the low natural frequency is the low structural rigidity of the whole milling head. Different from the integral rotary milling head, the main body of the A/B biaxial rotary milling head we studied is divided into two parts: the spindle box and the support frame. The spindle box is connected to the support frame through the rotation center, and the meshing contact of the rack and pinion is adopted at the axis B swing system. Consequently, the overall structural rigidity of the milling head is low. The rigidity of the milling head can be improved by adjusting the parameters of the rack and pinion meshing device, and strengthening the connection between the spindle box and the support frame, so as to prevent the occurrence of resonance and the damage to the workpiece and the machine tool.

2.2. Modal test analysis

LMS Test.lab14A modal analysis system was used to carry out the modal test with original support. The method of single point excitation and multi-point measurement (SIMO) was used, and vibration pick-up points were selected according to the criteria in the study of Sun (2012) [4]. An example of point arrangement is shown in figure 3, and the selection of the excitation point is shown by the arrow in figure 4. The arrangement of vibration pick-up points on other surfaces of the milling head is similar. Simplify the whole assembly model in the Geometry, as shown in figure 5.

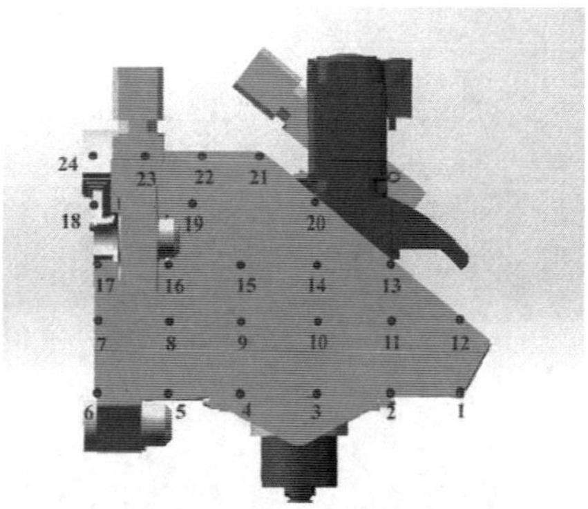

Figure 3. Layout of measuring points.

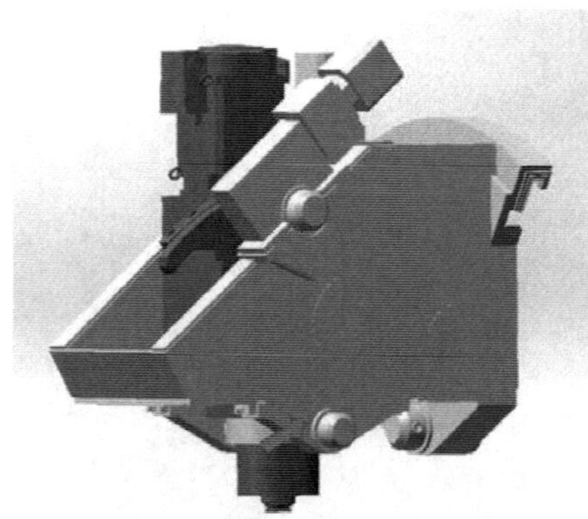

Figure 4. Location of excitation point.

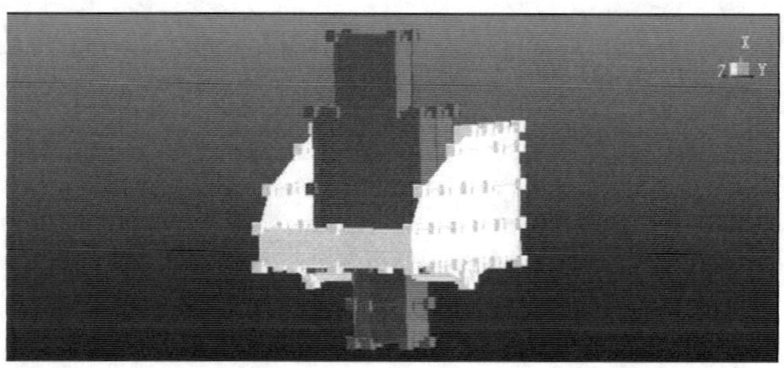

Figure 5. Simplified model of the milling head in LMS.

In the Modal Analysis, the Time MODF method was used to draw the steady-state diagram. Select the stable point at the peak of the frequency response function for modal extraction. The first six natural frequencies of the milling head are displayed in table 1.

2.3. Error analysis

The natural frequencies obtained by modal test analysis are compared with those obtained by finite element analysis, and the results are shown in table 1.

The maximum relative error of the frequency is 14.3%. Except for the first two natural frequencies, the relative errors of the other frequencies are all within 10%. The software ANSYS Workbench has some shortcomings, such as inaccurate modelling and mesh division, which will lead to some error of the simulation. The finite element model is validated by modal test within an acceptable error range [5].

Table 1. Comparison of the results between modal test analysis and finite element analysis.

Order	Frequency by modal test analysis/Hz	Frequency by finite element analysis/Hz	Relative error of the frequency/%	Modal shape
1	42.062	36.053	14.3	The support frame and the box rotate around axis Z.
2	50.091	56.145	12.1	The Box and the axis B system rotate around axis X.
3	76.858	79.693	3.7	The box does not move, and the support frame rotates around the axis X.
4	86.613	89.109	2.9	Take the plane YOZ as the symmetry plane, swing symmetrically left and right.
5	95.190	96.781	1.7	The milling head rotates along the axis Z as a whole.
6	98.331	104.140	5.9	The box extends up and down along the axis X.

3. Thermal error analysis of the A/B biaxial rotary milling head

In the working process of the milling head, the heat sources from outside and inside make the temperature rise of each part of the milling head uneven, leading to the thermal deformation of each part, which will reduce the long-term working characteristics significantly. The purpose of thermal error analysis is to calculate the specific situation of the thermal deformation of the milling head when it works in various heat sources for a long time. And the effects of the thermal deformation on the system accuracy are also taken into consideration, so as to guide the structural optimization design and the thermal error compensation of control system. After calculating the transient temperature field of the milling head accurately, the thermal deformation can be reduced and the machining accuracy can be improved by modifying the relevant structural parameters of the milling head or taking cooling measures.

3.1. Thermal transient simulation analysis of the milling head

In ANSYS Workbench, the heat flux coefficient and the heat transfer coefficient calculated were taken as thermal load conditions, and the thermal transient analysis was carried out to obtain the temperature and deformation values of the milling head at different times. In order to observe the change of the temperature field of spindle parts conveniently, the other parts except spindle and its corresponding parts were set transparent, and the change of the temperature field of the milling head and spindle was extracted. Take the temperature values of all measuring points (the locations of the measuring points are shown in figure 6) every 1800 seconds for recording, as is shown in table 2.

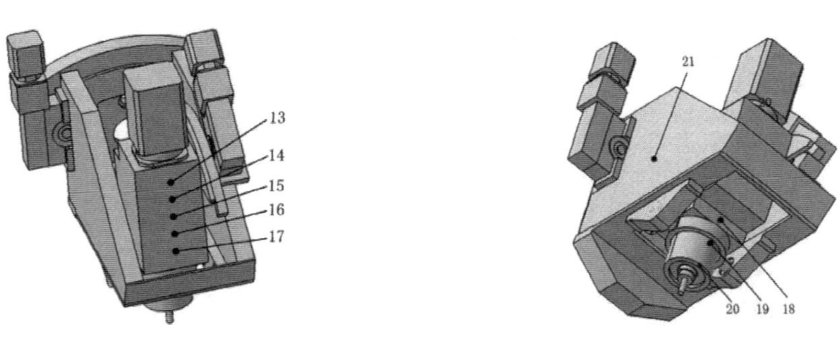

Figure 6. location of external test points.

Use the transient therm-structure coupling analysis module to solve the tip displacement at different times, and extract the tip displacement values in X, Y and Z directions and summarize them into table 3.

3.2. Thermal experiment analysis of the milling head

In order to verify the correctness of the theoretical analysis, the temperature rise under the actual working condition of the milling head was measured. By setting the acquisition software, temperature data were collected every 600 seconds, and displacement data were collected every second. 21 points for temperature measuring were arranged in total [6], among which 10 bearings and 2 gears were defined as measuring points 1-12. The specific positions of the other measuring points were shown in figure 6. The points 13-17 were on the outer surface of the spindle box. One point was arranged every 120mm. The points 15, 18 and 19 corresponded to bearings. The point 20 corresponded to the cutter head and the point 21 corresponded to the surface of the support frame.

3.3. Comparative analysis between experimental results and simulation results

3.3.1. Comparative analysis of temperature results. Collect the temperature values of measuring points at bearing 7, gear 1, cutter head and box surface 1 (i.e. measuring point 15, whose position is opposite to the heat source bearing). Compare the simulation results with the experimental results, analyse the error of the simulation results, and arrange the data as shown in table 2 below.

Table 2. Comparison between the temperature measured by simulation and measured by experiment.

Location of measuring point	Time/s	1800	3600	5400	7200	9000	10800
Bearing 7	Simulation value / °C	38.4	44	46.2	47.8	48	48.2
	Experimental value / °C	35.6	42	44.5	45.4	46	46.4
	Error / °C	2.8	2	1.7	2.4	2	1.8
Gear 1	Simulation value / °C	29.3	34.2	34.5	34.8	35.5	36
	Experimental value / °C	25.5	31.7	32.5	33.2	33	32.5
	Error / °C	3.8	2.5	2	1.6	2.5	3.5
Box surface 1	Simulation value / °C	28.4	32.7	36.4	37.3	38.4	39.6
	Experimental value / °C	24.1	34.5	34.9	35.2	35.8	36.3
	Error / °C	4.3	-1.8	1.5	2.1	2.6	3.3
Cutter head	Simulation value / °C	36.3	42.6	44.8	45.6	45.8	45.6
	Experimental value / °C	32.5	40.7	41.9	42.8	43.1	43.7
	Error / °C	3.8	1.9	2.9	2.8	2.7	1.9

It can be seen from table 2 that the simulation value is generally higher than the experimental value as to temperature. And the error is different at different times, among which the error at 1800s is the largest. Nonetheless all the error is within 5℃. Observing the temperature comparison of all the measuring points, these rules are still valid, and the error values are generally concentrated at 2°C-4°C. It can be found that the rising trend of the experiment results is similar to that of the simulation results through scatter. The results above show that the temperature simulation results are effective and the error is acceptable in the engineering practice.

3.3.2. Comparative analysis of displacement results. The simulation displacement is compared with the experimental displacement at the corresponding time, and the error is calculated, as shown in table 3.

Table 3. Comparison between the displacement measured by simulation and the displacement measured by experiment.

Direction	Time/s	1800	3600	5400	7200	9000	10800
X	Simulation value /mm	0.0038	0.0066	0.0082	0.0082	0.0085	0.0087
	Experimental value /mm	0.0028	0.0048	0.0067	0.0077	0.0082	0.0081
	Relative error	36%	37.5%	22.3%	6.5%	3.7%	7.4%
Y	Simulation value /mm	0.0039	0.0075	0.0103	0.0119	0.0122	0.0129
	Experimental value /mm	0.0037	0.0066	0.0091	0.0107	0.0115	0.0118
	Relative error	5.4%	13.6%	13.2%	11.2%	6.1%	9.3%
Z	Simulation value /mm	0.0217	0.0438	0.0630	0.0736	0.0776	0.0778
	Experimental value /mm	0.0237	0.0406	0.0565	0.0672	0.0705	0.0703
	Relative error	-9.2%	7.9%	11.5%	9.5%	10.1%	10.7%

It can be seen from table 3 that the relative error between the simulation value and the experimental value of the thermal deformation displacement in each direction of the tool tip is quite large, and it even reaches more than 30% in direction X when the time reaches 1800s and 3600s. It tends to be stable after 7200s and is within 10% in direction X. The symmetrical structure of the milling head results in small deformation in direction X in the experiment, while the simulation results of displacement are larger due to the simplification of the model and other reasons. In directions Y and Z, all the relative error of the displacement at different times is within 15%. It can be seen through scatter that the trends of simulation and experimental results are basically the same, and the simulation values of the displacement at the tool tip are generally higher than the experimental ones, which is consistent with the analysis of temperature.

4. Conclusion

From the simulation and experiment results we can conclude several important points:

1.In this paper, a kind of A/B biaxial rotary milling head is taken as the research object. Through the modal analysis of ANSYS Workbench, the first six natural frequencies and modal shapes of the milling head are obtained. The first six frequencies of the milling head are also measured by LMS Test.Lab. The relative error between simulation and experiment is almost acceptable. The first two frequencies of the milling head is relatively low, so the rigidity of the milling head can be improved to prevent the resonance of the milling head working under the alternating load, resulting in deformation and damage.

2.The transient thermal model and the transient therm-structure coupling model of the milling head system are established by using the finite element method. In order to verify the accuracy of the model, temperature and displacement measurements are carried out on the key points of the milling head under the actual processing conditions. The error of temperature and displacement is acceptable. The data of tool tip displacement obtained from the experiment can be of great help to the research of thermal error compensation of this kind of milling head.

3.The finite element model of modal analysis and thermal error analysis can truly reflect the working condition of the milling head system. Analyse and optimize the dynamic characteristics of milling head through the finite element method, then reduce the machining error caused by chatter and thermal deformation, so as to improve the machining accuracy and the level of CNC machine tools.

Acknowledgement
The authors would like to thank the National Science and Technology Major Special Project (China) *High grade CNC machine tools and basic manufacturing equip*ment (2016ZX04003001), January 2016 to December 2019.

References
[1] Zhang K, 2010 *Structural optimization and dynamic analysis on A/C biaxial rotary CNC universal milling head,* Master dissertation, Harbin Institute of Technology
[2] Bao X, 2012, *Structure finite element analysis of direct-drive A/C axis bi-rotary CNC universal milling head,* Master dissertation, Shenyang University of Technology
[3] Wang S, 2016, *Modal analysis of woodworking swing milling head based on hammering method,* Manufacturing Automation. **38(02)** 89-91+110
[4] Sun J, 2012 *The dynamic characteristic research of milling and grinding machine key parts,* Master dissertation, Nanjing Agricultural University
[5] Qin X, Lu C, Wang Q, Li H and Gui L, 2012, *Modal analysis of helical milling unit,* Advanced Materials Research. **482-484** 2454-2459
[6] Liu Y, Chen X, Yang Q and Zhang S, 2019, *Experimental method research of CNC machine tool thermal characteristics,* Machine Tool & Hydraulics. **47(03)** 74-78

A predicting method for the health state of the rolling bearing

Hongyan Jiang [a,b,c,*], **Dianjun Fang** [a,c]

[a] Tongji University, ShangHai 201804, China
[b] Jiangsu University of Science and Technology, ZhenJiang 212003, China
[c] Sino-German Institute for Intelligent Technologies and PhD Study, QingDao 266000, China

*E-mail: 1733316@tongji.edu.cn

Abstract. As an important component of rotating machinery, the health of rolling bearings is related to the normal operation of the whole machinery. A Period-Sequential Index (PSI) method was used for forecasting the three working states of the rolling bearing: health, fault, or failure. A sample from the vibration signals of the bearing, running only 2538 min because of a fault of its inner race, were tested by using the PSI algorithm. The results show that, this proposed PSI algorithm showed satisfactory forecasting quality in predicting whether the bearing is working in a healthy state or not. And, the value of the mean absolute percentage error at a fault state or a failure state was higher than that at a health state.

1. Introduction

Bearing failure often occurred in rotating machinery [1,2], which could result in breakdown of facility as well as costly downtime. So bearing monitoring was a big challenge of machine maintenance and automation processes. Especially in a big data era, a large number of time-serie datasets can be continuously generated in the bearing monitoring system [3,4], which has received much interest in predicting the health state of the bearing. Now, growing numbers of new algorithms have been proposed to forecast time series, such as exponential smoothing (ES), moving average (MA), support vector machine (SVM) and auto-regressive integrated moving average (ARIMA) method, and so on [5-9]. However, no traditional methods are suitable for the prediction of the bearing health. In the study, a new Period-Sequential Index (PSI) algorithm with four factors, which are period index, sequential index, small period index, super sequential index, is firstly introduced to predict the health state of the running rolling bearing.

2. Theoretical model

Finding Index-values carrying implicitly structure information usable, a heuristic method, the Period-Sequential Index (PSI) method [10], is used to predict the serial time-series data set. Fig.1 shows the schematic diagram of the PSI algorithm. H_{-2}, H_{-1} denote historical periods, i.e. the year before last and last year. H_0 represents the forecasting period. $PI(t_i)$, $SI(t_i)$, $pi(t_i)$, $si(t_i)$ describe the period index, sequential index, small period index and super sequential index, respectively.

2.1 Period index PI(t_i)

The period index indicates the relationship between the historical data and the reference value. The period index can be explained through equation (1) and (2),

$$PI_{-2}(t_i) = \frac{y_{-2}(t_i)}{K_{-2}} \tag{1}$$

$$PI_{-1}(t_i) = \frac{y_{-1}(t_i)}{K_{-1}} \tag{2}$$

where $y_{-2}(t_i)$ and $y_{-1}(t_i)$ describe the historical data at time of t_i during the period of H_{-2} and H_{-1}, respectively. K_{-2} and K_{-1} are reference functions of period index. A standard period average is originally set to be a reference function of period index, and where it is defined as a constant.

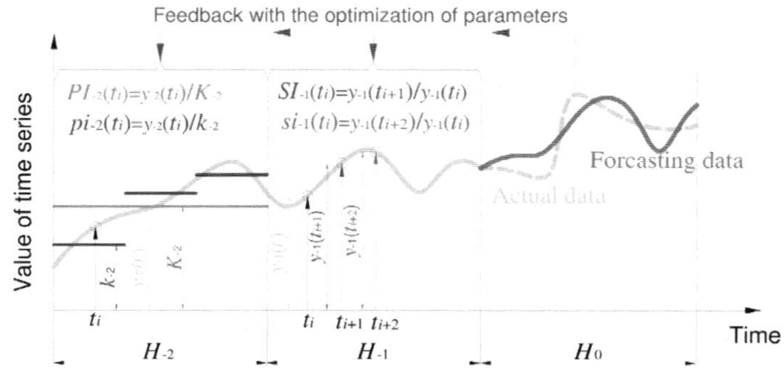

Figure 1. The schematic diagram of the PSI algorithm.

2.2 Sequential index $SI(t_i)$

The sequential index indicates the relationship between two adjacent historical data with the defined time steps. It is calculated through expression (3) and (4).

$$SI_{-2}(t_i) = \frac{y_{-2}(t_{i+1})}{y_{-2}(t_i)} \tag{3}$$

$$SI_{-1}(t_i) = \frac{y_{-1}(t_{i+1})}{y_{-1}(t_i)} \tag{4}$$

2.3 Small period index $pi(t_i)$

The small period index indicates the relationship between the historical data and the reference value, which can be explained through equation (5) and (6),

$$pi_{-2}(t_i) = \frac{y_{-2}(t_i)}{k_{-2}} \tag{5}$$

$$pi_{-1}(t_i) = \frac{y_{-1}(t_i)}{k_{-1}} \tag{6}$$

where k_{-1} and k_{-2} are reference functions of small period index. Here a small period (such as three months) average, because of seasonal factors, is originally set to be a reference function of small period index.

2.4 Super sequential index $si(t_i)$

The super sequential index indicates the relationship between two interval historical data with the defined steps. It is calculated through expression (7) and (8).

$$si_{-2}(t_i) = \frac{y_{-2}(t_{i+2})}{y_{-2}(t_i)} \tag{7}$$

$$si_{-1}(t_i) = \frac{y_{-1}(t_{i+2})}{y_{-1}(t_i)} \tag{8}$$

Therefore, the forecasting value at time of t_i during H_0 period is given by $F(t_i)$

$$F(t_i) = c \cdot PI(t_i) \cdot K_0 + d \cdot SI(t_{i-1}) \cdot y_0(t_{i-1}) + e \cdot pi(t_i) \cdot k_0 + f \cdot si(t_{i-2}) \cdot y_0(t_{i-2}) \tag{9}$$

where,

$$PI(t_i) = a.PI_{-2}(t_i) + b.PI_{-1}(t_i) \tag{10}$$

$$SI\ (t_{i-1}) = a.SI_{-2}(t_{i-1}) + b.SI_{-1}(t_{i-1}) \quad (11)$$

$$pi(t_i) = a.pi_{-2}(t_i) + b.pi_{-1}(t_i) \quad (12)$$

$$si(t_{i-2}) = a.si_{-2}(t_{i-2}) + b.si_{-1}(t_{i-2}) \quad (13)$$

$$K_0 = \gamma_0 \cdot K_{-1}\ ;\ k_0 = \gamma_0 \cdot k_{-1} \quad (14)$$

$$y_0(t_{i-1}) = \gamma_0 \cdot y_{-1}(t_{i-1});\ \ y_0(t_{i-2}) = \gamma_0 \cdot y_{-1}(t_{i-2}) \quad (15)$$

where, *c, d, e, f* (=0.25) are the weighing factors of forecasting equation (9). K_0 and k_0 are the correction coefficients for period index and small period index, respectively. *a,b* (=0.5) are the weighing factors of period index in H_{-2}, H_{-1} or the factors for sequential index. γ_0 is planning factor. The mean absolute percentage error (*MAPE*), given in the below formula (16), is used to evaluate the obtained results.

$$MAPE = \frac{1}{N}\sum_{i=1}^{N}\left|\frac{y_0(t_i) - F(t_i)}{y_0(t_i)}\right| \times 100\% \quad (16)$$

where $y_0(t_i)$ is the measured value at time of t_i during H_0; $F(t_i)$ is the forecasting value at time of t_i during H_0. *N* represents the number of model forecasting samples.

3. Simulation results and discussions

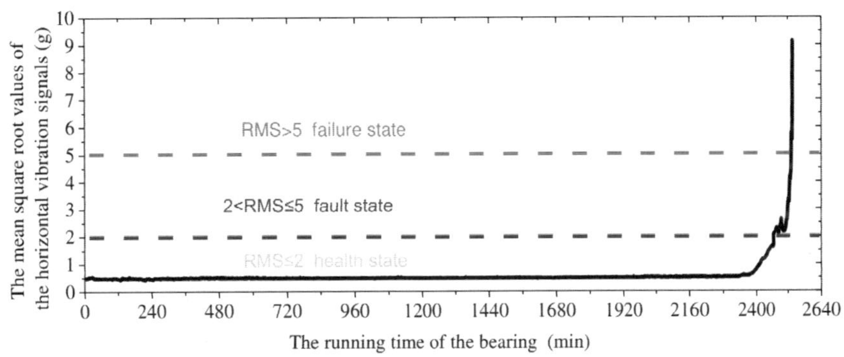

Figure 2. The mean square root (RMS) values of the bearing horizontal signals.

A sample [11] from the horizontal vibration signals of the rolling bearing were used and trained to forecast the three working states [12]: health (RMS≤2), fault (2<RMS≤5), and failure (RMS>5). In this sample, the rolling bearing with 10 kN load was operated at a speed of 2400 rpm, and run only 2538 min because of a fault of its inner race. The sampling frequency and period were set to 25.6 kHz and 1 min, respectively. Fig.2 showed the mean square root values of the bearing horizontal signals. Then, using the equation (9), simulation results were obtained as follows.

Fig.3, Fig.4 and Fig.5 showed the bearing horizontal signals and prediction in different working states. First, after training the historical datasets from 480 min to 540 min, the RMS values from 540 min to 570 min were predicted by using the PSI algorithm, as shown in Fig.3. It was seen from Fig.3 that, the predicted RMS values were all less than 2 g, and very close to the observed values, and a lower *MAPE* value of 1.61% was achieved by Eq. (16). The forecasting results also showed that, the bearing worked at a health state now between 540 min and 570 min. Then, the periodic detection results from 2400 to 2460 min and the prediction results between 2460 and 2490 min, were shown in Fig.4. It was seen from Fig.4 that, the real value and the predicted value reached 2.0 g at the running time of 2465 min and 2469 min, respectively. Although a time-lag of 4 minutes, the fault state of the bearing was still quickly predicted by using the PSI method. Finally, the simulation results in the working failure state was further shown in Fig.5. The predicted value was just more than 5.0 g at the running time of 2533 min, which nearly was equivalent to the real failure time of 2532 min. However, the higher *MAPE* values of 4.42%

IOP Conf. Series: Materials Science and Engineering 1063 (2021) 012005 doi:10.1088/1757-899X/1063/1/012005

and 6.26% were also achieved by Eq. (16) when the bearing worked in a fault state and failure state, respectively. The example showed that, the proposed PSI algorithm achieved a high accuracy in predicting whether the bearing was working in a healthy state or not.

4. Conclusions

(1) A Period-Sequential Index (PSI) method with four factors, which are period index, sequential index, small period index and super sequential index, was applied for forecasting the three working states of the rolling bearing: health, fault, or failure.

(2) During a health state, the predicted RMS values were less than 2 g and close to the observed values. In a failure state, the predicted value was just more than 5.0 g at the running time of 2533 min, which nearly was equivalent to the real failure time of 2532 min. The proposed PSI method showed satisfactory forecasting quality in predicting whether the bearing is working in a healthy state or not.

(3) The lower *MAPE* value of 1.61% was achieved in the health state; while the higher *MAPE* values of 4.42% and 6.26% were achieved when the bearing worked in a fault state and failure state, respectively. So the *MAPE* value at a fault state or a failure state was higher than that at a health state.

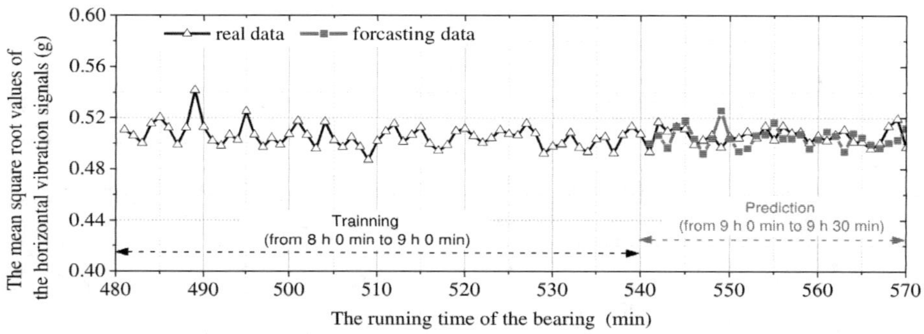

Figure 3. The bearing horizontal signals and prediction in health state.

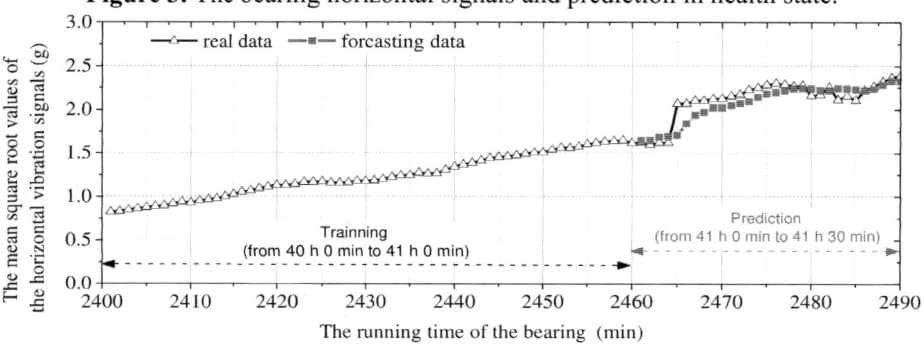

Figure 4. The bearing horizontal signals and prediction in fault state.

IOP Conf. Series: Materials Science and Engineering 1063 (2021) 012005 doi:10.1088/1757-899X/1063/1/012005

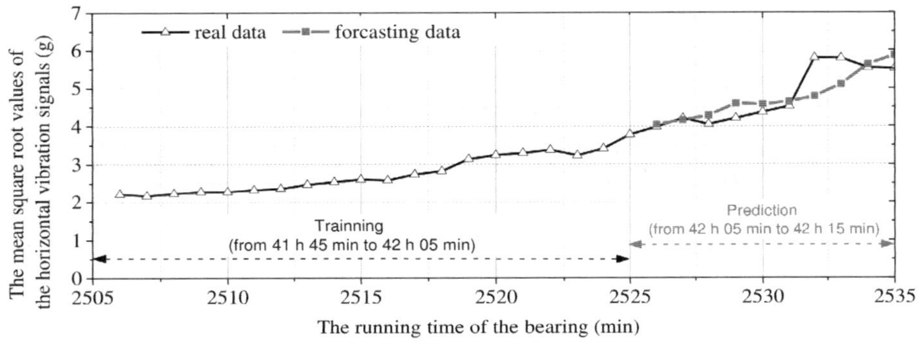

Figure 5. The bearing horizontal signals and prediction in failure state.

Nomenclature

a, b : weighing factors of period index or sequential index
c, d, e, f : weighing factors of forecasting equation
$F(t_i)$: forecasting value at time of ti during H_0 period
$G()$: measurement equation for forecasting value
H_{-2}, H_{-1}: historical period
H_0: forecasting period
-2, -1, 0 : subscripts, and represent during period of H_{-2}, H_{-1}, H_0, respectively
k_0 : correction coefficients for small period index
k_{-1}, k_{-2} : reference function of small period index
K_0 : correction coefficients for period index
K_{-1}, K_{-2}: reference function of period index
MAPE: mean absolute percentage error
N : number of forecasting samples
$pi(t_i)$: small period index at time of t_i
$PI(t_i)$: period index at time of t_i
$si(t_i)$: super sequential index at time of t_i
$SI(t_i)$: sequential index at time of t_i
t_i : time point
$y_0(t_i)$: measured value at time of t_i during H_0 period
$y_{-1}(t_i)$: measured value at time of t_i during H_{-1} period
$y_{-2}(t_i)$: measured value at time of t_i during H_{-2} period
γ_0 : planning factor

Acknowledgment

The authors gratefully acknowledge the support of key projects for international cooperation in scientific and technological innovation between governments, through grant No. 2017YFE0101600.

References

[1] Huang R.; Xi L.; Li X., Liu C., Qiu H., Lee J. Residual life predictions for ball bearings based on self-organizing map and back propagation neural network methods. Mechanical Systems and Signal Processing 2007, 21: 193–207.
[2] Narendiranath Babu T.; Devendiran, S; Aravind Arun; Rakesh A.; Jahzan M. Fault Diagnosis on Journal Bearing Using Empirical Mode Decomposition. Materials Today: Proceedings, 2018, 5:12993–13002.
[3] Wang Z.; Du W.; Wang J. Research and Application of Improved Adaptive MOMEDA Fault Diagnosis Method, Measurement. 2019, 140: 63-75.
[4] Wang Z.; He G; Du W.; Zhou J.; Han X.; Wang J.; He H.; Guo Xi.; Wang J.; Kou Y. Application of Parameter Optimized Variational Mode Decomposition Method in Fault Diagnosis of Gearbox.

IEEE Access. 2019, 7:44871-44882.
[5] Billah, B.; King, M.L.; Snyder, R.D.; Koehler, A.B. Exponential smoothing model selection for forecasting. International Journal of Forecasting, 2006, 22: 239-247.
[6] Barrow, D.K. Forecasting intraday call arrivals using the seasonal moving average method. Journal of Business Research, 2016, 69: 6088-6096.
[7] Stul, M.; Stul, K.; Leenders, R.; Butaye, L. Development of a SVM Prediction Model to Optimize the Energy Consumption of Industrial Installations by Detecting and Classifying Errors at an Early S. International Journal of Mechanical Engineering and Robotics Research, 2017, 6: 108-113.
[8] Eltayb, N. S. M.; Hamdy, A. LS-SVM Approach for Predicting Frictional Performance of Industrial Brake Pad Materials. International Journal of Mechanical Engineering and Robotics Research, 2018, 7: 105-112.
[9] Büyüksahin, U.C.; Ertekin S. Improving forecasting accuracy of time series data using a new ARIMA-ANN hybrid method and empirical mode decomposition. Neurocomputing, 2019, 361:151-163.
[10] Jiang H.; Fang D.; Spicher K.; Cheng F; Li B. A new period-sequential index forecasting algorithm for time series data. Applied sciences, 2019, 9: 4386: 1-12.
[11] Xi 'an Jiaotong University-Sun Rising Science and Technology Joint Laboratory-the experimental data of accelerated life of bearings, 2020, https://www.sumyoungtech.com.cn/news/xianjiaoda-yangkejilianheshiyanshifabuzhouchengjiasushoumingshiyanshujuzhulinindephmyanjiu.html.
[12] Li Y.; Ma X.; Zhao Y. A condition-based maintenance model for a three-state system subject to degradation and environmental shocks. Computers & Industrial Engineering, 2017, 105: 210-222.

IOP Conf. Series: Materials Science and Engineering 1063 (2021) 012006 doi:10.1088/1757-899X/1063/1/012006

High Stock Removal Hard Turning of Hardened 52100 Steel with High-Pressure Coolant Applied

Xiaozhong Song*, Roman Dreussi and Rahul G. Chaudhari

The Timken Company, Manufacturing R&D, North Canton, OH 44720, USA

*E-mail: xiaozhong.song@Timken.com

Abstract. Hard turning experiments were conducted on hardened 52100 steel utilizing high-pressure (10.3 MPa) coolant. CBN inserts were used. The cutting conditions were investigated to obtain high stock removal with a feasible tool life on the large part hard turning process. A small thermocouple was mounted in a ground slot underneath the CBN insert tip, close to the cutting edge, to effectively measure the insert's temperature during cutting. The cooling effect of the high-pressure coolant was evaluated by monitoring CBN insert crater wear and the time for the edge to become prematurely fractured. The temperature measurements show that the high-pressure coolant cools the insert more efficiently and has better lubrication capability leading to longer cut time without edge fracture. The study revealed that conventional low-pressure (0.4 MPa) coolant could not economically achieve high stock removal hard tuning due to high cutting temperature leading to premature fracture of the cutting edge. When using the high-pressure coolant and the optimized parameters, the tool life was enhanced by nearly 100%. The machined surface microstructure did not degrade even with the generation of very high shearing and friction heat energy in the cutting zone. Based on validation cut results and production part cycle assessment, the application of high-pressure coolant enables feasibility of high stock removal in hard turning with the ability to improve the productivity of large part rough stage turning by about 46%.

1. Introduction

52100 steel is widely used in the manufacturing industry because it can easily and cost-effectively be through hardened during heat treatment to get the necessary or desired hardness. After hardening and tempering, the material's microstructure transforms from austenite to tempered martensite and can attain hardness up to 58–62 HRC. Compared to carburized hardened microstructures, through-hardened tempered martensitic microstructures contain relatively large size, high-density, hard carbide particles. During hard turning of 52100 hardened steel, these hard carbide particles act very abrasively, like a grinding wheel's abrasive grains, and can produce the same scratch effect on the tool and chip interfaces as grinding did. The friction heat generated at the tool and chip interface when they slide against each other under high contact pressure is one of the largest contributors to the acceleration of the cutting tool's edge failure [1].

Over the authors' many years of experience with hard turning practices, they have seen that 52100 hardened steel is much more difficult to machine because cutting tool wear occurs so quickly, making its application in large part hard turning less economical or not feasible, since this reason the grinding is still dominative even the hard turning has the advantages of quick setup, process very flexible. The goal of the study presented in this paper is to examine ways in which 52100 steel's hard turning capability can be improved to expand its usefulness in large part hard turning. The authors propose the common-sense approach that could be achieved by applying high-pressure coolant to remove the

additional heat and enhance the cooling effects to lengthen tool life.
Pigott [2] was the first to consider this idea. He conducted tests using high-pressure coolant in steel turning with high-speed steel tools. He applied coolant at a pressure of 2.76 MPa directly on the flank face clearance. This cooling enhancement significantly increased the high-speed steel tool life. More recently, Mozammel Mia and Nikhil Ranjan Dhar [3,4] studied hard-turned EN-24T hardened steel using 80 bar high-pressure coolant and found that because the high-pressure coolant increased the cutting zone interface's cooling efficiency and lubrication, the machined material surface roughness was significantly improved, and tool wear was reduced to the point that a carbide insert could be used in hard turning applications. Rosemar B. da Silva et al. [5] assessed the tool life of PCD tools when turning Ti-6Al-4V alloy using conventional and high-pressure coolants. They found that tool life increased when coolant pressure was increased to 15 MPa. However, when pressure was increased beyond 15 MPa, the opposite trend was observed. They therefore proposed that an optimized coolant pressure region existed which was related to the total heat generated during the machining. Akira Hosokawa [6] recently studied turning in difficult-to-machine stainless steel with high-pressure coolant; he found that in lower cutting speed ranges, the high- pressure coolant had more cooling effect than in high cutting speed ranges, and the coolant's effectiveness in reducing chip breakage became significant only when the pressure was increased to over 7 MPa.
All these studies and investigations indicate that using high-pressure coolant to assist in machining can provide benefits such as longer tool life, higher machined surface quality and reduced chip breakage. However, all the published works were lab tests, and it has not been demonstrated in production that using high-pressure coolants can significantly extend the length of the high stock removal hard turning cutting time. Additionally, a high rate of stock removal requires longer tool life during cutting, but the conventional coolant pressure (0.4 MPa) cannot extend tool life sufficiently.
This paper presents an experimental investigation of high stock removal hard turning with high-pressure coolant applied. A small thermocouple was mounted in a ground slot underneath the CBN insert tip, close to the cutting edge, to measure the insert temperature in process to evaluate the cutting and cooling capability of the high-pressure coolant. Machining experiments were conducted using CBN inserts under different cutting conditions that led to a high stock material removal rate. The cooling capability was evaluated by monitoring cutting temperatures, tool life extension, the capability of chip breakage and process feasibility. Conventional low-pressure (0.4 MPa) coolant cutting was tested first as the benchmark for identification of the current process limit and tool failure mode, and then experiments on the three primary cutting parameters (speed, feed and depth of cut) were performed to establish optimization conditions. The optimization conditions were validated via the production parts cut. Under the high stock removal conditions, the 10.3 MPa coolant pressure setup increased tool life by about 100%. Based on this result, it was concluded that the large part hard-turn feature could be completed within the extended tool life (no tool failure during the test cut), and productivity could be expected to increase by 46%.

2. Experiment setups

2.1. Setup and data measurement equipment
From production consideration, machining experiments were carried out on through-hardened 52100 bearing-grade steel. Carburized hardened C8620 steel was also selected for the tool life comparison. The steels' main chemical compositions are shown in Table 1. The 52100 steel's hardness was heat-treated to the range of 58~62 HRC. The C8620 steel was carburized and hardened to the same surface hardness, with a hardened layer thickness of 1.5 mm to 3.0 mm. The sample workpiece was made from a part with an OD of 400 mm and an available cutting length of 50 mm. The tests were conducted with a high-precision CNC vertical lathe using a magnetic chuck to hold the workpiece on the face. All the test cuts were carried out on the OD straight surface.

Table 1. Chemical composition of test steels.

Harden steel	C %	Mn %	Cr %	Ni %	Mo %
52100	0.93~1.1	0.25~0.45	1.3~1.6	0.14~0.25	0.04~0.1
C8620	0.8~1.0	0.7~0.9	0.4~0.6	0.4~0.7	0.15~0.25

An industrial high-pressure pump with a pressure capability of up to 13.8 MPa and a rated maximum flow rate of 5 L/min was used. Pressures ranging from 8.0 to 10.3 MPa were tested. A Sandvik standard tool holder equipped with a high-pressure coolant jet nozzle applied the coolant. The cutting tool used was a Sumitomo CNGA turning insert with a tool nose radius of 1.2 mm and a CBN tip grade of BNC200. The temperature measurement thermocouple was made from K type resistant wire with a thermal sensing conjunction tip size of about 0.2 mm. The thermocouple was mounted in a ground slot underneath the CBN insert tip. The distance from the tip location to the cutting edge was about 0.8 mm perpendicular to the flank face. The output temperature analog signal from the thermocouple was acquired by a Yokogawa DL750 digital waveform recorder with a built-in thermal voltage signal amplifier module. The coolant used in the experiments was water-based. Figure 1 shows the experimental machine, tooling, thermocouple mounting location and high-pressure coolant setup.

Figure 1. High-pressure coolant hard turning experiment setup: (a) high-pressure coolant connection to the tool, (b) thermocouple installation location, (c) coolant ejection status.

2.2. Experiment procedures

The hard-turning experiments were conducted with the following conventional low pressure and high-pressure coolant parameters.

- The first set of experiments was conducted to cut material and measure the temperature difference between the conventional (0.4 MPa) and high-pressure coolant conditions. The purpose of this test was to validate that the high-pressure coolant reduced cutting temperatures and to identify what effect it had on hard-turning the 52100 steel.
- The second set of experiments was conducted with the conventional low-pressure coolant applied at a higher material removal rate cutting to identify the tool failure mode under high material removal rate cutting conditions.
- The third set of experiments compared the tool life improvement results under the high-pressure coolant and validated the high-pressure coolant as the key factor in affecting cooling and lubrication capability and thereby improving tool life.
- The final test used the highest removal rate parameters identified in the previous tests and production parts to validate the feasibility of successfully finishing the production cut and assessing whether productivity had been enhanced.

3. Results of experiments and discussion

3.1. Temperature under the conventional and high-pressure coolant conditions

All the temperature data collected from the experiments under different cutting conditions are shown in Figure 2. Figure 2(a) shows the cutting temperature measured during hard turning using lower stock removal with the conventional coolant pressure (0.4 MPa) applied.

The highest temperature was about 130 Celsius degrees at the measurement location, which was 0.8 mm underneath of the insert tip rake face. Although this was not the direct cutting point surface temperature, it was proportional to it, enabling the quantitative evaluation of the rake face temperatures under the different turning process conditions. Figures 2(b), (c) and (d) present the turning temperatures at high stock removal rates with conventional coolant and high-pressure coolant supply conditions.

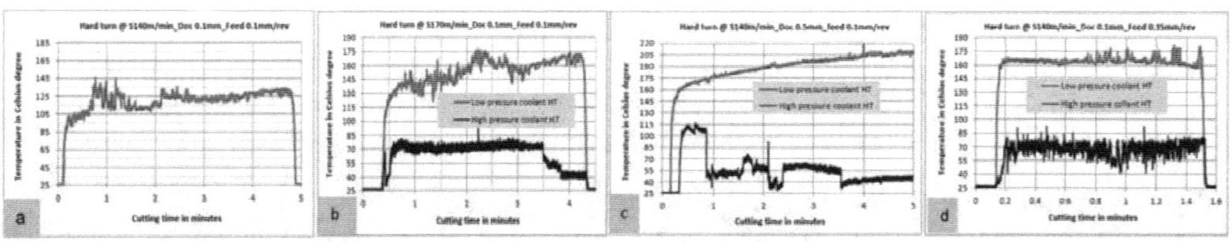

Figure 2. Temperature comparisons of high-pressure (10.3 MPa) coolant vs. conventional (0.4 MPa) coolant during hard turning at different process cutting conditions.

All above results clearly indicated that the high-pressure coolant application would have much lower cutting-induced temperatures than the conventional-pressure coolant application. The applied high-pressure coolant tests demonstrated significant cutting zone temperature reduction capability which gave the potential for the tool life to increase in high stock removal cutting.

The relationship of temperatures and material removal rates were further analyzed for each setup of conventional pressure coolant conditions. As shown in Figure 3, the increases in cutting speed, feed rate and depth of cut would increase the cutting material removal rate, however, the insert rake face temperature also increased, it shows the cutting speed has the most significant impact on the temperature change in the above three cutting parameter conditions.

Further examination of this data shows that at a fixed cutting depth Doc of 0.1 mm and feed of 0.1 mm/rev, as in Figure 3(a), the ratio of temperature increase to material removal increase since speed change is about 0.15°C/mm^3 (calculated from the data (170°C-125°C)/(1700mm^3-1400mm^3)). From figures 3(b) and (c), these ratios are 0.013°C/mm^3 and 0.011°C/mm^3, respectively. These results demonstrate that the temperature in the cutting zone is more sensitive to cutting speed than to the other two primary process parameters. So, from the process optimization point of view, with cooling condition fixed, the selection of a larger cutting depth or faster feed rate is the most effective strategy to achieve less tool wear by temperature reduction and in turn to gain economical productivity enhancement.

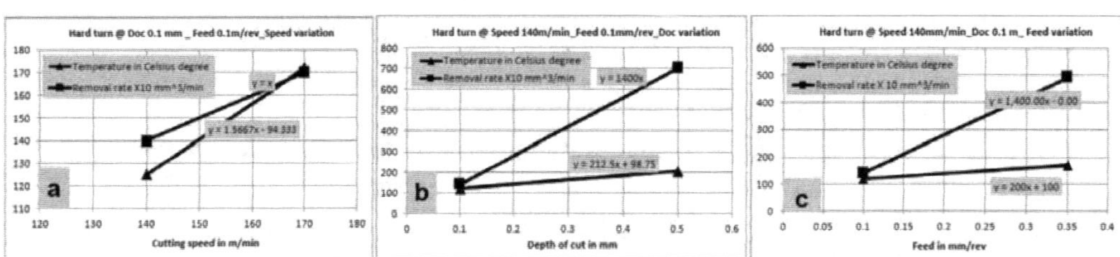

Figure 3. Hard turn material removal rate vs. average temperature increase under different setup conditions at low pressure coolant cutting application.

3.2. High stock removal hard turning with conventional coolant pressure and tool failure mode

Figure 4 shows images of insert wear and failure status on the rake and flank faces after the high stock material removal hard turning experiment in which the conventional-pressure (0.4 MPa) coolant was applied.

In Figure 4(a), the cutting tool crater wear is very severe. The flank wear, however appears to follow a normal pattern. As Figure 4(b) shows, its maximum flank wear land VB was less than 0.2 mm [7-8]. From the rough turning point of view, this amount of flank wear would be negligible. Although there was severe crater wear on the rake face, the insert cutting time was short, only 12 minutes, a much shorter length of time than normal production hard turning on the 52100 hardened steel case.

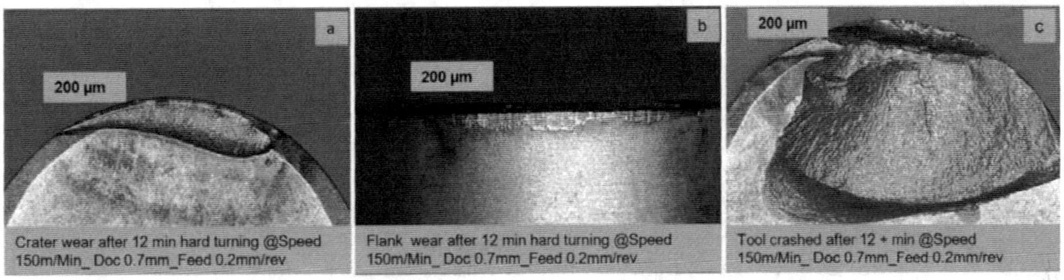

Figure 4. Tool wear after 12 minutes of hard turning at high stock material removal rate with conventional-pressure (0.4 MPa) coolant supply. (a) is crater wear at 12 minutes cut time, (b) is the flank wear and the (c) is the fracture happened between 12-16 minutes cutting time.

In the test after the tool wear was measured, the insert was put back for continued cutting, but the test could not be completed, the insert fractured between the 12 minutes to 16 minutes portion of the test cycle as Figure 4(c) shows, it can be observed that the insert edge fractured from the highest-temperature point at the cutting edge lead-in side, which is also the maximum crater wear location. From this significant rake face wear/fracture characteristics, it can be concluded that under the high stock material removal cutting condition, tool damage is due primarily to rapid rake face crater wear and sudden transition to cutting-edge fracture.

To further understand what failure mode caused the insert to suddenly fracture, a crater wear test was conducted numerous times at a relatively lower cutting speed (100 m/min instead of 150 m/min), utilizing the high material removal rate cutting condition setup. This time, the insert was observed under a high-magnification microscope at four-minute check intervals in the test cycle. As shown in Figure 5, one deep micro crack and material spalling spot was observed in the crater area, which initiated from the highest-temperature spot after the insert had been in use for 24 minutes of cutting.

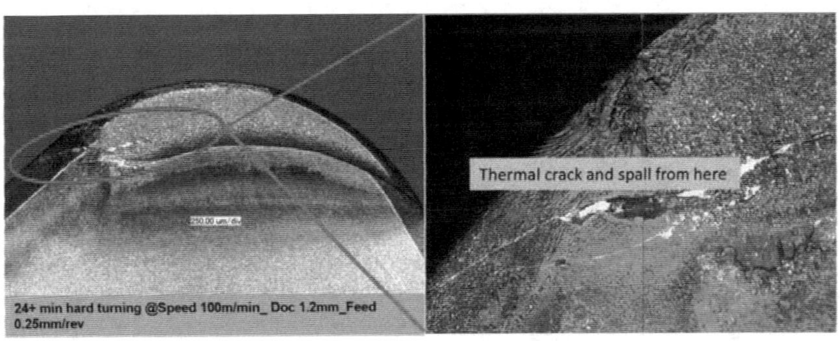

Figure 5. Tool wear after 24 minutes of hard turning at high stock material removal rate with conventional-pressure (0.4 Mpa) coolant supply.

IOP Conf. Series: Materials Science and Engineering 1063 (2021) 012006 doi:10.1088/1757-899X/1063/1/012006

It could be assumed that during cutting, the insert rake face experienced mechanical abrasive wear and, simultaneously, a thermal chemical oxidization reaction occurred on the tool rake face. Figure 6 is an SEM photo of an insert crater wear spot on the rake face after hard turning. The black dots are CBN grains and the gray area is bonding material. An EDS analysis was performed at spot 1 and 2 locations (one CBN grain and one bonding spot). The EDS analysis results indicated that high oxygen content was detected in the spot 2 location. In almost all cases, the metal oxidation structure was brittle and weaker in strength than the initial bonding metal [9].

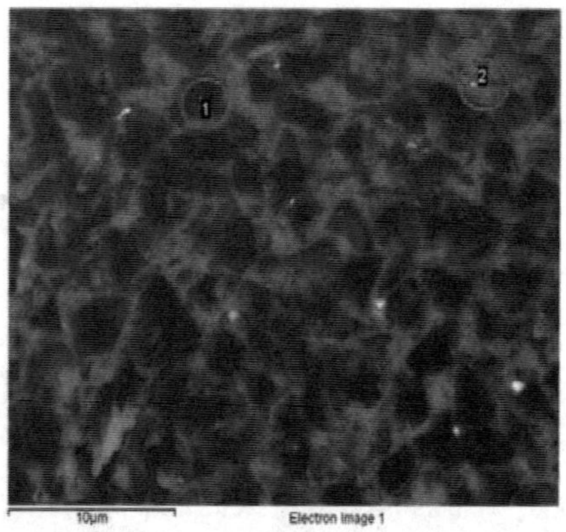

Spectrum	In stats.	B	C	N	O	Al	Ti	Total
1	Yes	28.01	17.93	45.44	3.63	1.21	3.77	100.00
2	Yes		8.09	31.11	20.26	5.08	35.46	100.00
Max.		28.01	17.93	45.44	20.26	5.08	35.46	
Min.		28.01	8.09	31.11	3.63	1.21	3.77	

Figure 6. Chemical components found in crater wear area via EDS analysis.

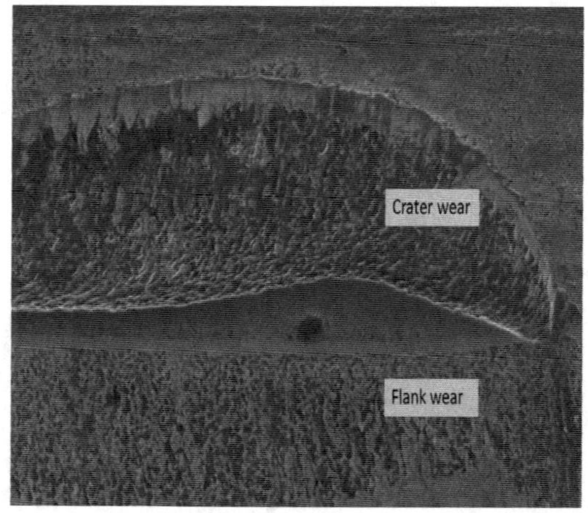

Figure 7. Asperity surface structure in insert wear area after hard turning.

The spot 1 location possessed a relatively higher concentration of boron elements and a much lower oxygen component. The spot 1 and 2 comparison indicated the insert bonding metal material had undergone a strong chemical oxidization reaction under high temperature. This oxidization reaction directly weakened the tool bonding material strength and was then wiped away by the flowing chip. The insert rake face wear process was accelerated by the hard carbide particles in the hardened 52100 steel chips. Figure 7 shows a typical used insert wear pattern after hard turning. Its surface appears to have an asperity layer that was formed by the CBN grains and oxidization of the bonding metal, with some adhesive steel buildup from the workpiece.

When the crater wear on the cutting edge became deep enough, an area of thermal stress concentration developed. When this concentrated stress exceeded the surrounding bond metal's strength limit, it generated a thermal crack. As the experimental cutting and wear continued, the cutting edge became more and more weak, and the crack propagated quickly into the neighboring area. Impelled by the thermal and cutting force impact, the crack extension finally induced the bulk bonding metal to fracture. In other words, insert failure was dependent on how fast the crater wear developed.

3.3. Cooling capability and tool life improvement after application of high-pressure coolant

The above description of the benchmarked (conventional coolant pressure) tool failure mode shows that improvement in tool fracture/failure performance in high stock removal conditions with conventional coolant pressure could be achieved except for improving the strength of insert material's thermal mechanical performance, which is most insert manufacturers' current development priority. Using high-pressure coolant to accelerate the reduction of heat created in the cutting zone is an effective alternative method

The images in Figure 8 compare the chip thermal effect and size breakage capability of hard turning experiments on the 52100 through-hardened steel.

Figure 8. Chip formation comparison under conventional and high-pressure coolant conditions.

The chips formed in Figure 8(a) under the conventional coolant pressure condition have a long strap shape. Since the contact length between the chip back face and insert rake face is longer, as the heat generation model in Figure 9 shows, this longer contact length creates the potential for more friction and heat to be introduced from the sliding interface area. Conversely, in Figure 8(b), all the chips formed under the high-pressure coolant condition are in the shape of short pieces. Further, when comparing the chip colors in Figures 8(a) and (b), one can see a clear difference: Under the conventional coolant condition, the chips are a dark color, which indicates that the steel experienced a strong oxidation reaction that occurred in the high-temperature environments. But under the high-pressure coolant condition, the chips have retained their original metal color. This characteristic coloration indicates that the high-pressure coolant reduced the high-temperature chemical oxidation reaction in the cutting zone not just by reducing the chip and insert rake face contact length, but also by producing a hydrodynamic lift effect that further reduced contact friction and carried away more heat from the shear plane.

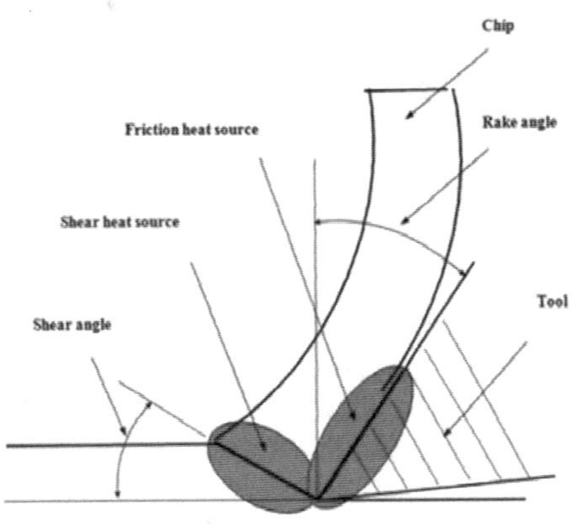

Figure 9. Heat source model of turning process. The main heat is from material shearing, chip back face's touch creating scratch friction effect in the material removal process.

Figure 10 is a tool crater wear evolution comparison of high stock material removal hard turning under the different coolant pressure application conditions. The comparison images clearly demonstrate that in the high-pressure coolant case, it takes nearly twice as long before tool crater wear developed as it does in the conventional low-pressure coolant case (54 minutes vs. 27 minutes). Longer tool life means larger dimension parts for which the hard-turning process can be carried out with no tool fractured. In production the tool life should be long enough to finish at least one single cycle pass with no mid-cycle tool change. It should be noted that to meet the one complete pass tool life restriction in production practice the material removal rate (lower feed, depth of cut, etc.) must be sacrificed to achieve longer tool life.

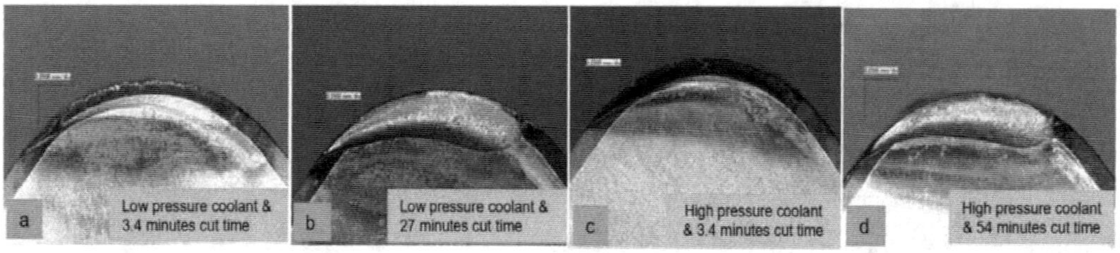

Figure 10. Tool crater wear evolution comparison at cutting conditions: speed 100 m/min, feed 0.2 mm/rev, and DOC 1.0 mm.

3.4. Cutting speed and material effect on the hard turn machinability under high pressure coolant condition

In the previous temperature tests, it was demonstrated that the higher the material removal rate, the higher the cutting temperature will be. The temperature increase is most sensitive to the cutting speed. The next set of experiments was performed to validate that this principle applies to tool wear either in the lower or high-pressure coolant condition.

The images in Figure 11 show the results of tool wear experiments using different cutting speeds. It is clearly indicated that even under the high-pressure coolant condition, tool crater wear is still very sensitive to cutting speed. These results are consistent with the measured temperature test results. With regards to tool life, it can also be seen that increasing material removal rate by increasing speed is the last option for process optimization.

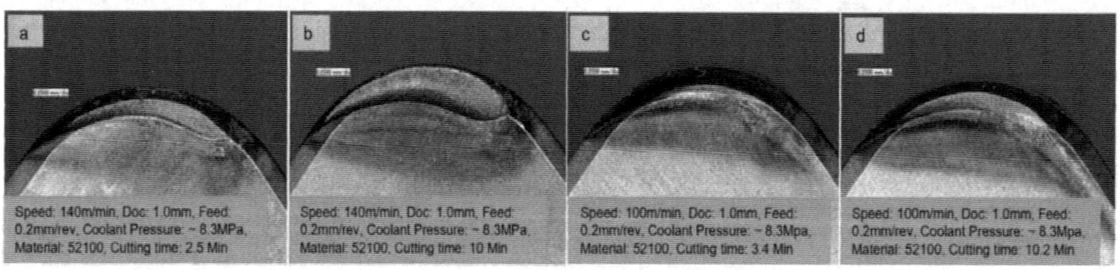

Figure 11. Comparison of cutting speed effect on tool crater wear under high pressure coolant condition.

The images in Figure 12(a)(b)(c)(d) are of carburized C8620 hardened steel tested under the high stock removal rate and high-pressure coolant conditions and comparison with through harden steel 52100. The images indicate that when same cutting parameters are set up with regard to cutting time, faster feed and larger depth of cut, the carburized steel manifests much less tool crater wear. The authors thought except the high pressure coolant effect since carburized steel has less hard carbide particle scratch effect that benefits the tool wear reduction. It is almost four times less when crater area is used as the evlaution criteria.

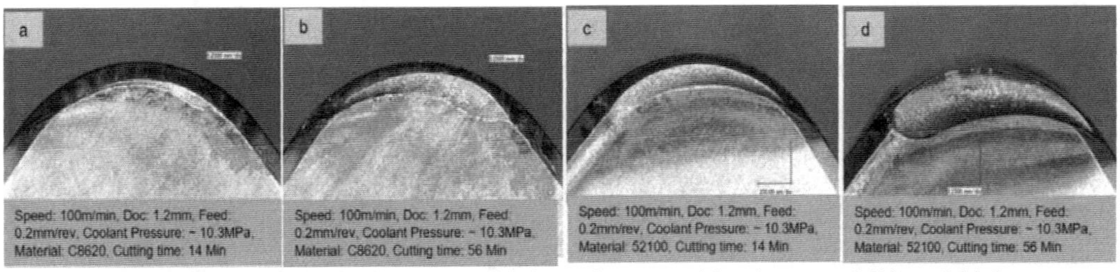

Figure 12. Workpiece material effect under high stock removal hard turn tool life and high-pressure coolant conditions.

3.5. Workpiece material microstructure change under high pressure coolant condition

In Figure 13, it could be seen how material surface microstructure is affected after high stock removal hard turning with the high-pressure coolant. On the top of the machined surface, a mechanical hardened layer on the order of 1–2 μm can be seen. Judging by the color under this layer, no further thermal damage has occurred. If the layer had undergone thermal damage, it would be in the tempered black color. Since the high stock removal process is used for rough and semi-finishing, even if a thin mechanical hardened surface layer develops, it would be removed by the finishing step, which is normally set up at depth of cut of 100 μm.

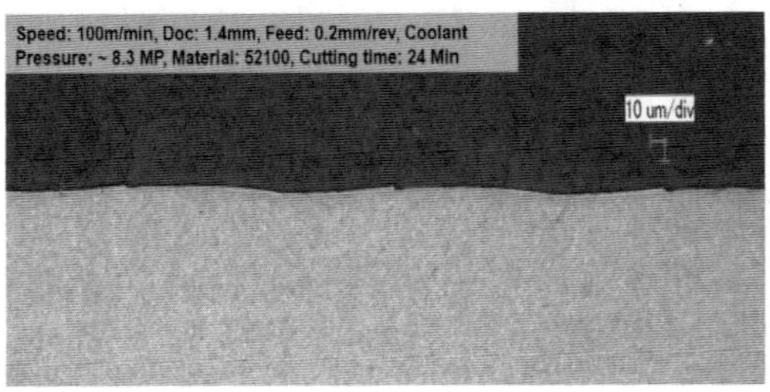

Figure 13. Surface microstructure under high stock removal and high-pressure coolant conditions.

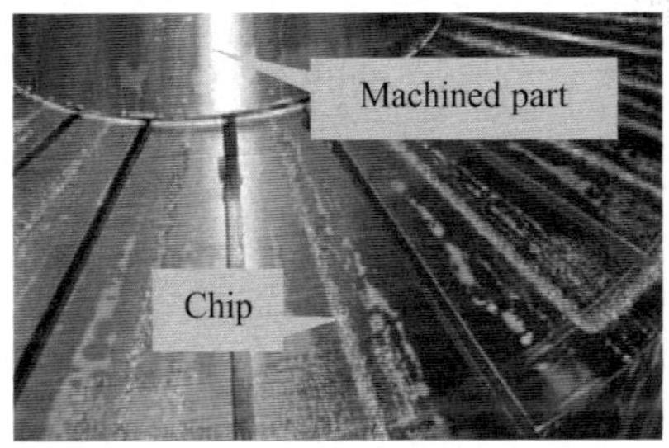

Figure 14. 52100 hardened steel part validation cut and chip control status after validation cut.

4. Large diameter part validation cut and productivity enhancement estimation

As shown in Figure 14 machined part, one 52100 hardened steel part was selected to conduct the validation cut at the high stock removal rate (30,000 mm^3/min setup) with surface speed 100 m/min, feed rate 0.25 mm/rev, DOC 1.2 mm, and the high-pressure 10.3MPa coolant applied.

Figure 14 not only shows that the cutting was completed successfully, but also clearly shows that the high-pressure coolant has very good chip control capability. Figure 14 showed validation cut is the new insert edge test case, no big chips remain on the chuck table, all the cut chips are very small and can easily be flooded away by the high pressure flushing coolant, only a few of them are held by the magnetic poles on the chuck table.

Another large part with 1.92 meter diameters was selected for the case study to estimate the productivity gains that could be achieved in practice with this method of high stock removal hard turning. The OD and ID hard turned surfaces were about 72 mm and 260mm long, Table 2 provides a comparison of the cycle time reduction if the high-pressure coolant is used in the process instead of the conventional low-pressure (0.4 MPa) coolant. Since it was feasible for this setup to apply the 30,000 mm^3/min material removal rate, the hard turn time was cut down from the initial 320 minutes to 171 minutes, which corresponds to a 46.5% cycle time savings in the process. This cycle time reduction represents a significant increase in productivity.

Table 2. Large cutting surface part cycle comparison

		Current LP & speed 182 m/min				Enhanced HP & speed 100 m/min			
HT features	**Stock (mm)**	**Doc (mm)**	**Feed (mm/rev)**	**Pass**	**HT time (min)**	**Doc (mm)**	**Feed (mm/rev)**	**Pass**	**HT time (min)**
OD	3.2	0.6	0.15	6	96	1.2	0.25	3	51
ID	2.4	0.6	0.15	4	224	1.2	0.25	2	120
		Total cutting time			**320**	**Total cutting time**			**171**

5. Conclusions

High-pressure coolant-applied high stock removal hard turning was tested in through-hardened 52100 and carburized hardened steel under different production parameters. The tool failure mode and proposed improvement approach were investigated with the goal of making the process feasible for large size product hard turning. Preventing the fast crater wear on the tool cutting edge was proved to be the key control factor. The following conclusions can be drawn from this investigation:

- Under high stock removal conditions, the quick tool failure is primarily due to the severe crater wear making the insert cutting edge weaker; at some point a thermal crack is induced, followed by sudden chipping and fractured out.
- In addition to removing more heat from the cutting zone, the high-pressure coolant jet enhances lubrication by creating a hydrodynamic fluid layer effect between the back face of the chip and the front of the tool rake face. It also breaks the long, continuous chips down to small, non-continuous pieces, which reduces the chip contact length in the tool rake face and consequently reduces friction/heat diffusion into the tool.
- With this enhanced cutting zone temperature reduction capability, tool life under the high stock removal condition could improve 100% in rough hard turning.
- The carburized hardened steel also has shown much better tool life under the high stock removal hard turning case since this material has less hard carbide particle scratch effect.
- The larger part production application case study indicates that productivity can be expected to increase by 46% when the high-pressure coolant is applied.

References

[1] J.A. Arsecularatne, L.C. Zhang, 2006 On Machining of Hardened AISI D2 Steel with PCBN Tools, *Journal of Materials Processing Technology*, Vol. **171**, 2006, pp 244–252

[2] D. Kramar, J. Kopac, 2009 High Performance Manufacturing Aspect of Hard-to-Machine Materials, *Advances in Production Engineering & Management*, Vol. **1–2**, April 2009, pp 3–14

[3] Mozammel Mia, Nikhil Ranjan Dhar, 2015 Effect of high pressure coolant jet on cutting temperature, tool wear and surface finish in turning hardened steel, *Journal of Mechanical Engineering,* Vol. **45**, June 2015, pp 1–6

[4] Mozammel Mia, Nikhil Ranjan Dhar, 2016 Prediction of surface roughness in hard turning under high-pressure coolant using artificial neural network, *Measurement*, Vol. **92**, 2016, pp 464–474

[5] Rosemar B. da Silva, Álisson R. Machado, 2013 Tool life and wear mechanisms in high speed machining of Ti–6Al–4V alloy with PCD tools under various coolant pressures, *Journal of Materials Processing Technology*, Vol. **213**, 2013, pp 1459–1464

[6] Akira Hosokawa, Yuki Nomura, Tomohiro Koyano, 2018 Turning of Difficult-to-Machine Materials with High Pressure Coolant, *17th International Conference on Precision Engineering*, Nov. 12–16, 2018, Kamakura, Japan, B-1-12

[7] ISO 3685:1993: Tool-life testing with single-point turning tools, (1993), 2nd edition.

[8] R. Serra, W. Rmili, 2016 Experimental Evaluation of Flank Wear in Dry Turning from Accelerometer Data, *International Journal of Acoustics and Vibration*, Vol. **21**, No. 1, 2016

[9] Xiaozhong Song, Rahul Chaudhari, Fukuo Hashimoto, 2014 Experimental Investigation of Vibratory Finishing Process, *Proceedings of the ASME 2014 International Manufacturing Science and Engineering Conference*, MSEC2014-4093, June 9–13, 2014, Detroit, Michigan, USA

Design, Test and FEM Analysis of Customized Titanium Alloy Implant with Scaffold Based on Additive Manufacturing

Wen-Teng WANG [1, a], Ruei-Nan SHENG [2,b] Zih-Liang JHUANG [3,c]

Dept. mechanical & computer-aided Engineering, National Formosa University, Taiwan

E-mail: [a]wtwang@nfu.edu.tw, [b]40271110@gm.nfu.edu.tw, [c]10820132@gm.nfu.edu.tw

Abstract. In order to design the prosthesis which can mimic the surroundings for bone cell proliferation and cell attachment and avoid stress shielding effect by matching the Young's modulus with different parts of human bone, we come up with the porous structure with scaffold design. This paper is presenting the scaffold of design, the mechanical property of the biomimetic scaffold and the relative error rate between the Finite element analysis (FEA) and compressive test. However, most of the models of the porous structure are designed in the STL format especially the high topological model which will make the FEM preprocessing procedure difficult. According to the reason above, we used the concept of the CT-based voxel mesh method for high complex STL model, which skips the time-consuming mesh and repairing reconstruction process. Lastly, we successfully designed the porous prosthesis according to the data from FEA and compressive test.

1. Introduction

In recent years, due to the rapid-aging society, the need of the prosthesis has increased, and the mature of the additive manufacture (AM) technology, the design of the prosthesis gets more flexible.

The three factors affecting the mechanical property of the implant are the types of unit cells of the scaffold, the layout of the scaffold, and the size of the pore, respectively. Through the algorithm aided design (AAD) platform, we can control three factors precisely, which makes the Young's Modulus of the implant match to the patient's bone defect area, lower the risk of the stress shielding effect [1].

As the mature of the AM technology, most of the model choose to be designed and saved in the STL format. However, this kind of format is required a complex preprocessing process for Finite Element Analysis (FEA) to repair the mesh and geometry reconstruction. The process is time-consuming and costs a lot of system resources, so most of the company decided to use trial and error to probe the Young's modulus of the scaffold. In order to increase the efficacy, in research we proposed a rapid method to help the complex STL model to skip mesh repair and geometry reconstruction process. Using the concept of voxelize mesh allows the complex STL model to execute directly the preprocess. After comparing the Young's modulus from FEA and compressive test, a stable error rate of 19% can be obtained. We can use the error rate as a reference for the design of the implant in the future.

2. The design of the scaffold

2.1. The Morphology of the Scaffold

Through the internal logic component of the AAD software "Grasshopper", we come out with three different types of unit cells. The first one also the basic one is the line structure-based scaffold, pore size, and the diameter of the struts are easy to control, but lack of topological trait. The second is the Gyroid scaffold, it has good topological trait and unanimous pore size, but the control flexibility of the pore size is very poor [2]. The last one is the biomimetic scaffold, this design was based on the Voronoi algorithm [3], using the randomly cell expanding concept, by mimicking the nature trabecular bone we can offer the similar surroundings for the bone cell to grow [4][5]. Three types of scaffolds are shown in the Figure.1(a) ,1(b), 1(c) pore size of these three types of scaffold are all above 300μm according to [5]'s research.

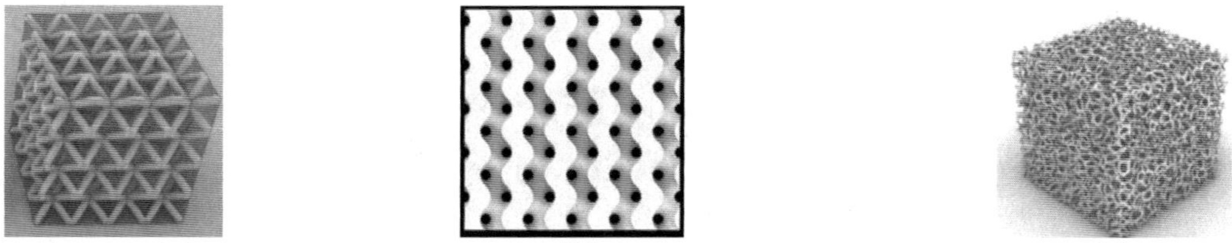

(a). **(b).** **(c).**

Figure 1. The Morphology of the Scaffold
(a) Line structure-based scaffold. (b) Gyroid scaffold (c) Biomimetic scaffold

2.2. The layout of the scaffold

In this study, we propose the method for the layout of the scaffold.

To start, we use the iso-curve to subdivide the iso-surface according to the UV value (Figure.2), and then the length and the width of the unit cell can be obtained after the subdivide. Next, we create two surfaces and then subdividing them with an equal amount of UV value in the three-dimensional region. Additionally, we extract the nodes from each surface partitions and connect these eight nodes to become a closed geometry.

In the end, we can split the unit cells along the Z-axis to define the desired height of the unit cell when the height of the unit cell has been obtained, as shown in Figure.3

Not like the other method we can't just define the centroid of the unit cell as a moving coordinate. The strut will not be contained inside the unit cell. In order to avoid this problem, we need to add the eight nodes of the unit cell into the moving coordinate, so that the strut of the scaffold will fit inside the unit cell. As shown in Figure.4.

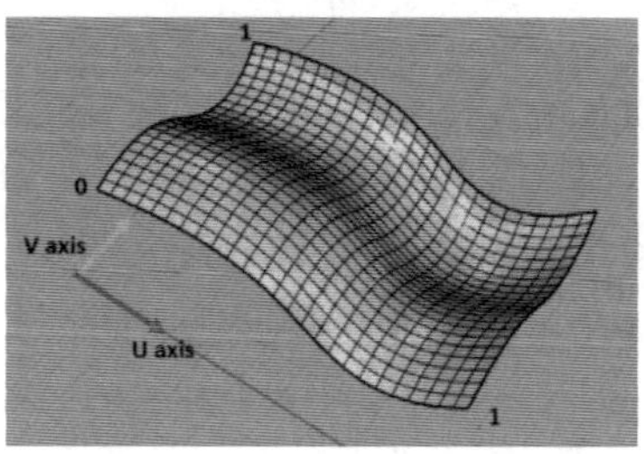

Figure 2. UV value of surface.

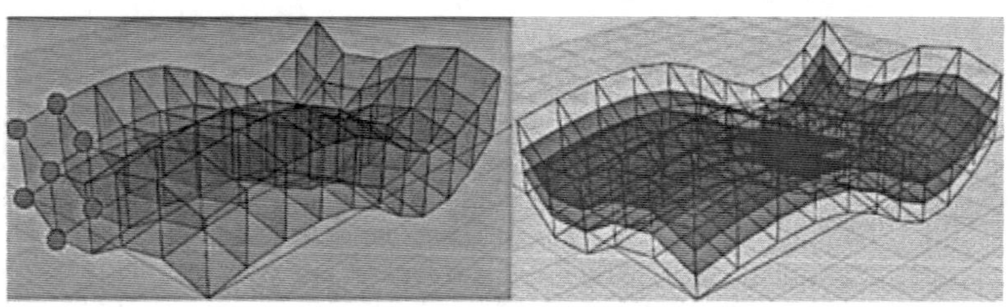

Figure 3. Unit cell of surface.

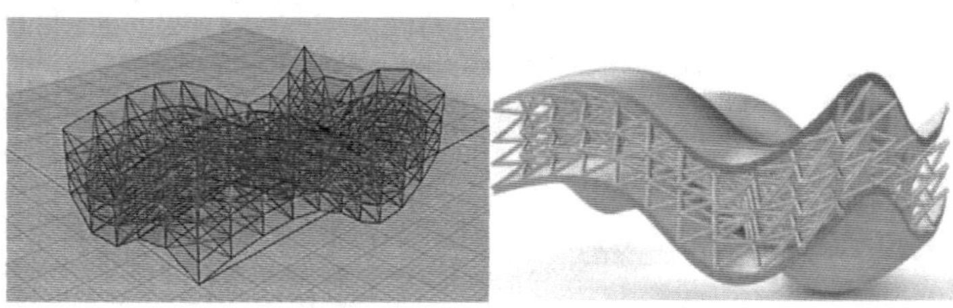

Figure 4. Iso-surface structure.

3. FEM analysis of the scaffold

3.1. Voxel Mesh & Boundary condition

The difference between the CT and the Stl-Based voxel mesh is that the CT-Based voxel mesh is generated directly from pixel to voxel, on the other hand, the Stl-Based voxel mesh consists of unsmooth hexahedron mesh (Figure.5), while the unsmooth hexahedron mesh doesn't make the significant effect on the result of stress and strain in FEA [6-8].

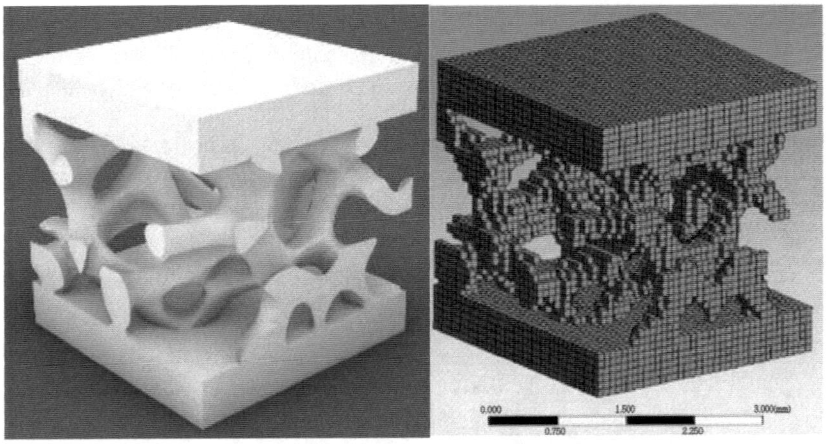

Figure 5. The unsmooth hexagonal mesh.

The purpose of this study is to explore the equivalent Young's modulus rather than the behavior beyond the elastic limit. As a result, we fix the bottom of the specimen, and make the displacement along the Z-axis in liner analysis, then calculate the Young's Modulus from the elastic gradient.

3.2. Mesh Convergence

In order to find out the most stable and efficient element size before proceeding with the FEM analysis, we will have a mesh convergence. The sizes of the elements are range from 0.15 to 0.06, as the size of the element become smaller, the cost of system resource will increase. As a consequence, we can clearly see that the deviation of the force became smaller, as shown in Figure.6. According to the cost of system resource and efficiency in research, we choose a 0.08mm element size to run the analysis.

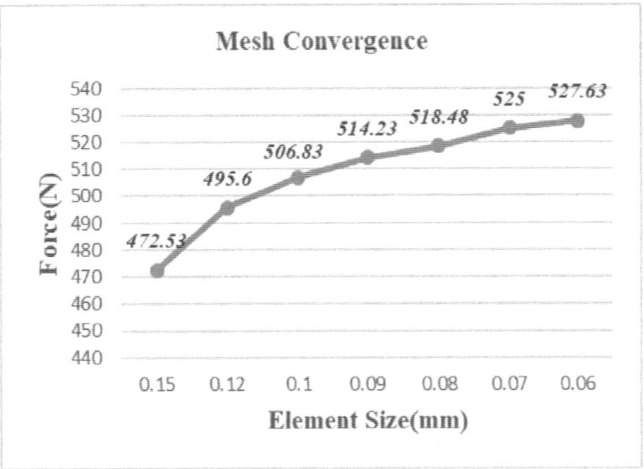

Figure 6. Mesh convergence.

4. Method and result of mechanical testing

In recent years most of the research still working on the geometry of the biomimetic scaffold, its mechanical property is rarely discussed. According to the reason, in this study, we follow the ISO13314 specification to conduct the compressive test. We use HUNG TA (HT-2402) computer servo-controlled material testing machine with 10^{-2} s^{-1} strain rate to test the eight different types of specimen of porosity from 80% to 54% (Figure7), while there are five specimens for each porosity,

The material property of the specimen can be obtained from the tensile test report offer by the specimen which is made by Renishaw metal 3D printing machine which uses selective laser melting (SLM) technique (Figure.8), due to the thickness of the metal powder and the power of the laser beam, the Young's Modulus of the SLM Ti6Al4V is 33.7Gpa [9].

According to the stress and strain data, we can obtain the Young's Modulus of each porosity, as shown in Figure.9. According to the Figure.9, we can discover the trend that the Young's Modulus will increase as the porosity decrease. However, there is one exception, the struts of the porosity 62%(B) were specifically design into a thin blade shape. As the result turns out, the blade shape affects the rigidity of the scaffold. According to the comparison between 62%(A)and(B) are shown in Figure.10, the mechanical property comparison is shown in table.1 [10][11].

Figure7. Porosity 80%、70% 、60.8%、54%.

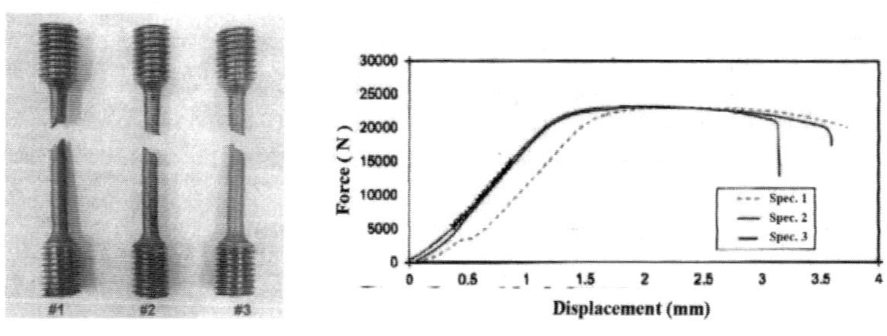

Figure 8. Tensile test.

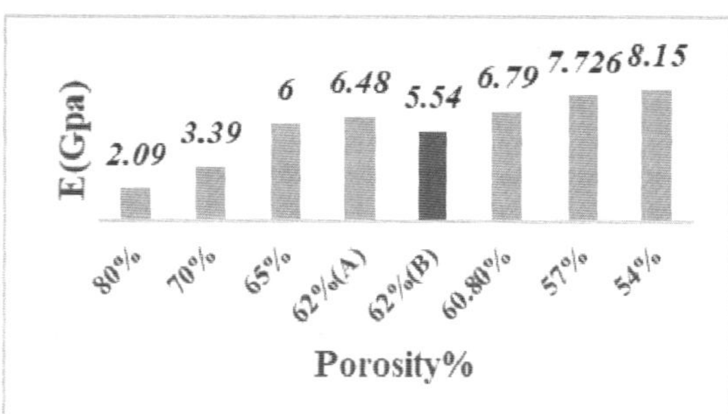

Figure 9. Young's Modulus of each porosity.

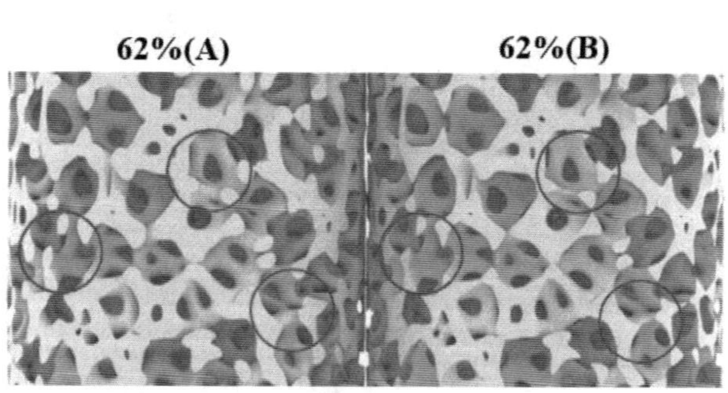

Figure 10. Structure of 62% (A) and (B).

Table 1. Mechanical property comparison.

Source	Type of bone	E(Gpa)	Porosity of the Scaffold
Runkle and Pugh(1975)	Human,Distal Femur	*8.69*	*54%*
Ku et al. (1987)	Human, tibia	*3.17*	*70%*
Mente and Lewis (1987)	Human, tibia	*5.3*	*62%(B)*
Choi et al. (1989)	Human, tibia	*4.59*	*62%(B)*
Kuhn et al. (1989)	Human, iliac crests	*3.81*	*70%*
Jensen et al. (1990)	Human, vertebra	*3.8*	*70%*
Amirouche et al. (2014)	Acetabular	*5.78~4.13*	*62%(B)*

5. Discussion

In order to facilitate the compare process, we choose four different porosity of the specimen for the FEM analysis. Comparing both of its Young's Modulus which obtain from compressive test and FEM analysis can realize that the exception of the error rate of the specimen P80 is 40% and others are under 19%, as shown in Figure.11. and Figure.12.

Turning now to the experimental result on the error rate, we can summarize the dimension inspection table (as shown in table.2). On the basis of table.2, we sort and make the error rate of the volume and length into the charts as shown in Figure.13. and Figure.14.

From the previous charts, it can be presumed that most of the error came from inside of the scaffold, which means there might be few powders that fail to exhaust still inside the scaffold. There are few powders inside the scaffold, which may cause the pore fail to take shape and become a closed pore indirectly increase the rigidity of the scaffold.

It is also worth noting that the specimen P80 has the highest volume error rate, the presumption explains the reason that the analysis error rate of P80 can reach 40%.

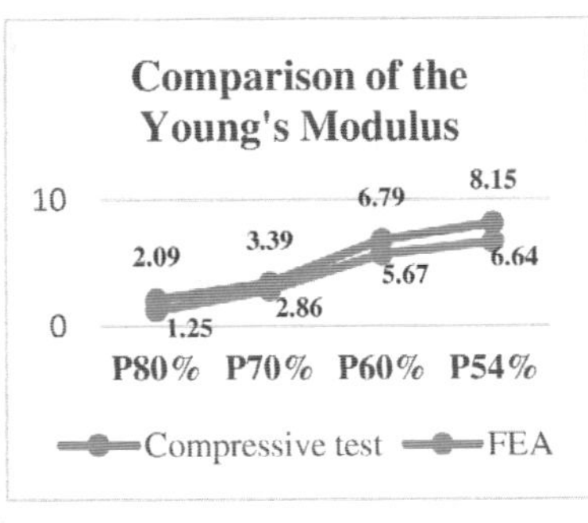

Figure 11. Comparison of the Yong's modulus.

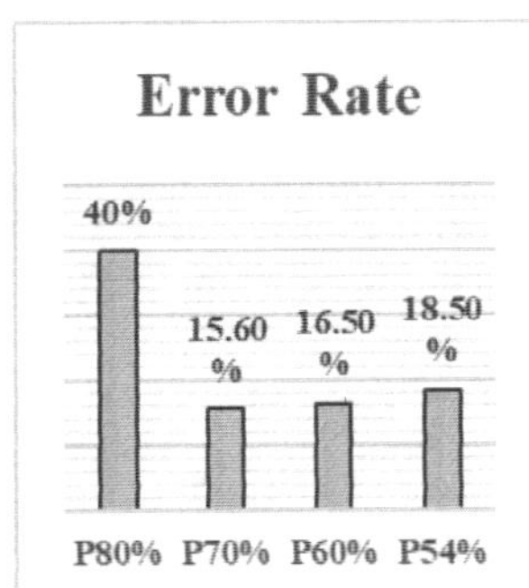

Figure 12. Error rate.

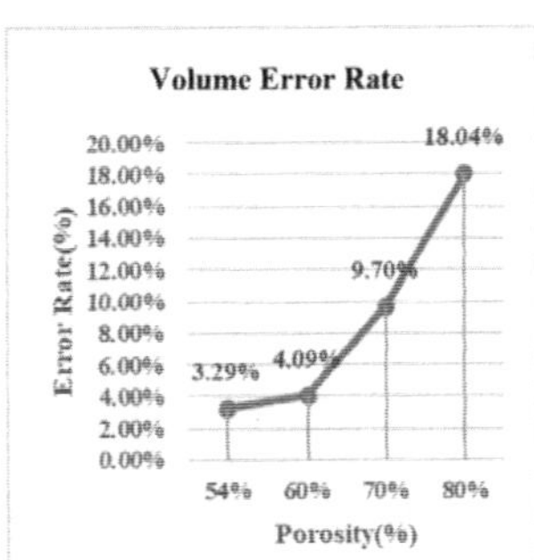

Figure 13. Volume error rate

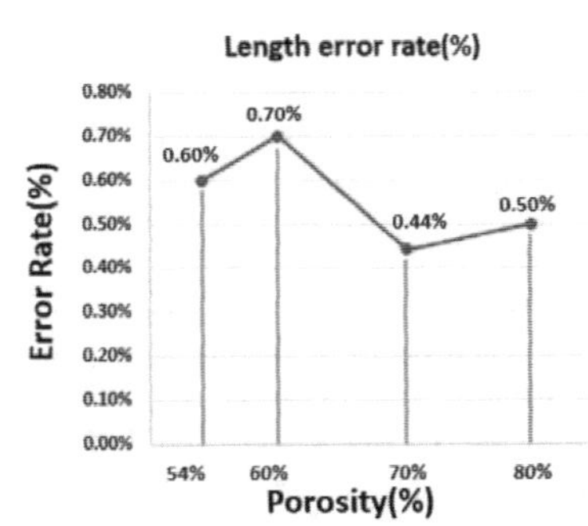

Figure 14. Length error rate

Table 2. Dimension inspection

	54%		60.8%		70%		80%	
	Length (mm)	Weight (g)	Length (mm)	Weight (g)	Length (mm)	Weight (g)	Length (mm)	Weight (g)
Standard	19	3.361	19	3.0174	19	2.425	19	1.864
	19.16	3.5	19.11	3.1	19.11	2.7	18.97	2.1
	19.02	3.4	19.11	3.1	18.99	2.6	19.16	2.1
	19.09	3.5	19.16	3.1	19.15	2.7	19.12	2.3
	19.12	3.5	19.16	3.2	19.11	2.7	19.13	2.3
	19.20	3.5	19.16	3.2	19.06	2.6	19.10	2.2

P.S.: The density is 4.417(g/cm^3)

In addition, we use the data of the result form FEA and compressive test to manufacture the porous prosthesis with similar mechanical properties to human bone, as shown in Figure15 and Figure16.

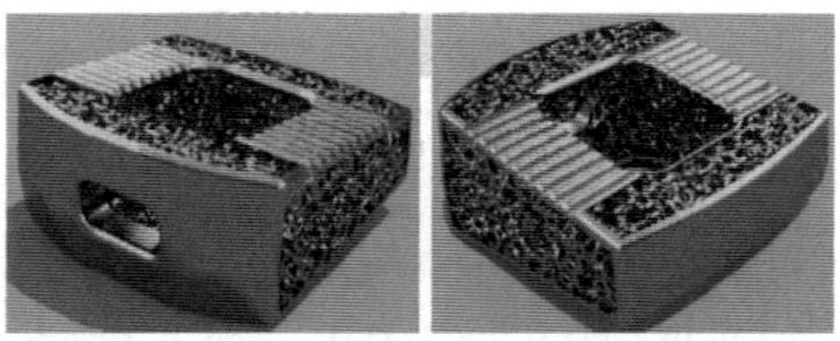

Figure 15. Fusion cage.

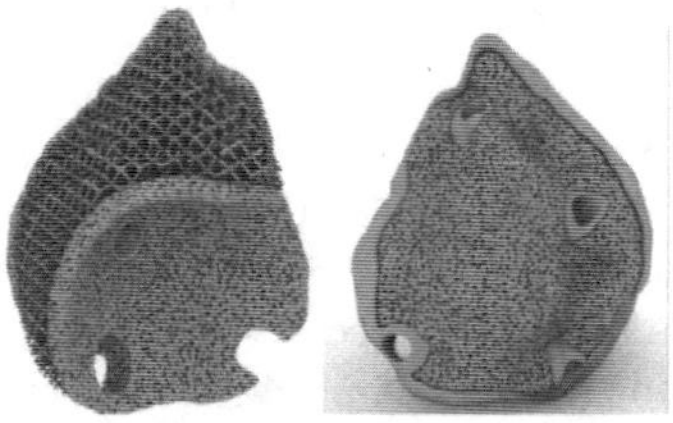

Figure16. Customized implant for hip.

6. Conclusion

In this study we successfully design the porous specimen with a scaffold, which can mimic the Young's modulus of trabecular bone, improve the chance of cell proliferation and cell attachment, etc. In the FEM analysis, we offer a more efficient method to replace trial and error, and we can obtain the most results of error rate under 19%.

However, there is the problem of precision for the AM machine. If the AM machine can achieve 40~90 μm, we can design and fabricate the structure of the compact bone and combine the trabecular bone, allow future design to be more flexible.

Acknowledgements

The authors gratefully acknowledge Metal Industries Research & Development Centre (Taiwan) for funding this research, and James Chen, vice president of Ingrowth Biotech for providing all of the samples made by metal additive manufacturing.

References

[1] M.I.Z. Ridzwan, Solehuddin Shuib, A.Y. Hassan, AA.Shokri and M.N.Mohamad Ibrahim, "*Problem of Stress Shielding and Improvement to the Hip Implant Designs*", Journal of Medical Sciences (2007)

[2] Scherer, M.R.J, "*Double Gyroid Structured Functional Materials Synthesis and applications*", ISBN:978 3 319 00353 5, (2013)

[3] Wojciech Pokojski, Paulina Pokojska , "*Voronoi diagrams inventor, method, applications*", Vol. 50, no. 3, pp. 141 150 DOI: 10.2478/pcr 2018 0009, (October 2018)

[4] M. Fantini, M. Curto & F. De Crescenzio, "*A method to design biomimetic scaffolds for bone tissue engineering based on Voronoi lattices*", ISSN: 1745-2759 (Print) 1745-2767 (Online), (2016)

[5] Vassilis Karageorgiou, DavidKaplan, "*Porosity of 3D biomaterial scaffolds and osteogenesis*", Biomaterials 26 5474–5491, (2005)

[6] J. H. Keyak, J. M. Meagher, H. B. Skinner and C. D. Mote, Jr, "*Automated three dimensional finite element modelling of bone: a new method*", J.Biomed.Eng, (1990)

[7] Thibaut Bardyn, Mauricio Reyes, Xabier Larrea, and Philippe Büchler, "*Influence of Smoothing on Voxel Based Mesh Accuracy in Micro Finite Element*", SPB 187800 Chapter ID 10, (2010)

[8] Rakesh Saxena and Tony S. Keller, "*Automatic Volume Mesh Generation Applied to Trabecular Bone, Department of Mechanical Engineering The University of Vermont*", Burlington, VT 05405, (1999)

[9] Yu-Lien Lin , "*Finite Element Simulation Analysis of Mechanical Properties of 3D Printing Periodic Element Structures*",Master's thesis, National Taipei University of Technology, Taiwan <https://hdl.handle.net/11296/tu67gj> , (2018)

[10] Jae Young Rho, Richard B.ashman and Charles H. Turner, "*Young's Modulus Of Trabecular and Cortical Bone Material: Ultrasonic and Microtensile Measurements*", J. Biomechanics Vol. 26, No. 2, Pp. 11i- 119 (1993)

[11] J. Biomechanics Vol. 26, No. 2, Pp. 11i- 11 9.Farid Amirouche, PhD, Giovanni F. Solitro, PhD, Mark Gonzalez, "*MD.Total Hip Arthroplasty Acetabular Components Stability and Management of Rim-Defect Position in Cementless Fit Technique*", ORS Annual Meeting, Poster No: 0960(2014)

IOP Conf. Series: Materials Science and Engineering 1063 (2021) 012008 doi:10.1088/1757-899X/1063/1/012008

Model of weld beads geometry produced on surface temperatures by Wire and Arc Additive Manufacturing (WAAM)

S. Manokruang[1*], F. Vignat[1], M. Museau[1] and M. Linousin[1]

[1]Univ. Grenoble Alpes, CNRS, Grenoble INP, G-SCOP, F-38000 Grenoble, France.

*E-mail: supasit.manokruang@grenoble-inp.fr

Abstract. Wire and Arc Additive Manufacturing (WAAM) is an Additive Manufacturing (AM) technology that knows a strong interest in recent years. With this technology, the combination of an electrical arc welding and a metallic wire feedstock is used to manufacture parts layer by layer. As part geometry depends strongly on deposition conditions, this study is focused on the characterization of single bead walls and their geometrical evolutions according to welding parameters and surface temperatures. Our experiment has consisted of the production of walls manufactured with a $5083Al$ wire and built from 1 layer up to 10 layers with several surface temperatures ($Tsur$ fixed at 100, 200, 300 and 400°C). Before the deposition of each new layer, we ensure that the substrate is at the correct temperature using a thermometer in order to guarantee constant welding conditions. Those welding parameters have been kept constant with a Wire Feed Speed (WFS) at 5 m/min and a Travel Speed (TS) at 0.6 m/min. Wall bead geometry and dimensions, width (w) and height (h), have been then measured using a 3D scanner. Results are used to feed a predictive model developed with a circular model curve fitting each top layer cross-section of walls. Data form this study, coupled with thermal simulation, will allow us to predict faithfully parts shapes and to control their dimensions.

1. Introduction

Additive Manufacturing (AM) is under great interest from the industry. The main idea of AM is to produce parts through deposition of material layer-by-layer instead of its removal. Within AM, various technologies and methods exist and are able to produce a metal structure [1]. Wire and arc additive manufacturing (WAAM) is one of them [2] and has the specificity to offer a high deposition rate with low energy [3]. However, during the deposition process, process parameters have to be kept constant in order to insure the geometrical stability of beads, width (w) and height (h) [4]. It is well-known that thermal accumulation is an important factor which greatly affect this stability as well as residual stresses [5]. In order to reduce this heat accumulation various solutions can be implemented [6]. Thus, Cold Metal Transfer (CMT) is a new technology welding process. CMT has been developed in various types and this research interested CMT-STD type. The CMT-STD is ideal for the welding of thin aluminum alloy sheets due to low heat input and little deformation. The welding electrical cycle divide into three phases: peak current phase, background current phase and short-circuiting phase [7].

This is the reason why the Cold Metal Transfer (CMT) technology have been designed for Al alloy thin plates and why it is particularly relevant for AM of aluminum parts [8-9]. Secondly, waiting time can be adjusted in order to reach a predefined and controlled temperature before starting a new deposit at a given place according to the desired temperature. However, heat accumulations can be heterogeneous, confined to specific parts zones (angles, beads crossing, huge section variations or geometric

singularities) where it could be difficult to predict, control and manage these temperatures. A possible solution to ensure optimal geometry quality for those regions could consist in adapting welding parameters such as Wire Feed Speed (WFS) and Travel Speed (TS).
This study details a first work conducted in order to understand geometrical evolutions of single wall beads according to welding parameters and a precise control of the surface temperature.

2. Equipment's and Methodology

2.1. Equipment's
Figure 1 shows the equipment's used for walls beads production, for instance a YASKAWA MA1440 Robot and a Fronius TPS CMT 4000 Advanced welding station (Fronius, Pettenach, Austria). For our experiment, an aluminum wire, 5356 alloy, with diameter of 1.2 mm has been used. Parts have been built on an ER5083 Al alloy substrate about 250 mm x 250 mm x 5 mm ($w * l * h$) which have been clamped during production. A working distance to 15 mm between torch and substrate has been fixed. An argon shielding gas was used with a flow rate of 13 L/min. Welding parameters have been kept constant for the whole experiment with Wire Feed Speed (WFS) and Travel Speed (TS) about 5 m/min and 0.60 m/min using the 875 CMT synergic law provided by Fronius. As we have to control the substrate temperature for the needs of our experiment, a heating device has been designed. Composed by 3 heating resistances (6 kW), it allows a temperature regulation up to 500°C thanks to an electrical controller. Thermocouples type k and thermometer (FLIKE-50S) are used to control and to check the surface temperature. A 3D laser scanner (Artec Space Spider), with a resolution of 0.10 mm, was used to achieve a precise numerical model of weld beads geometry from which width (w) and height (h) have been extracted.

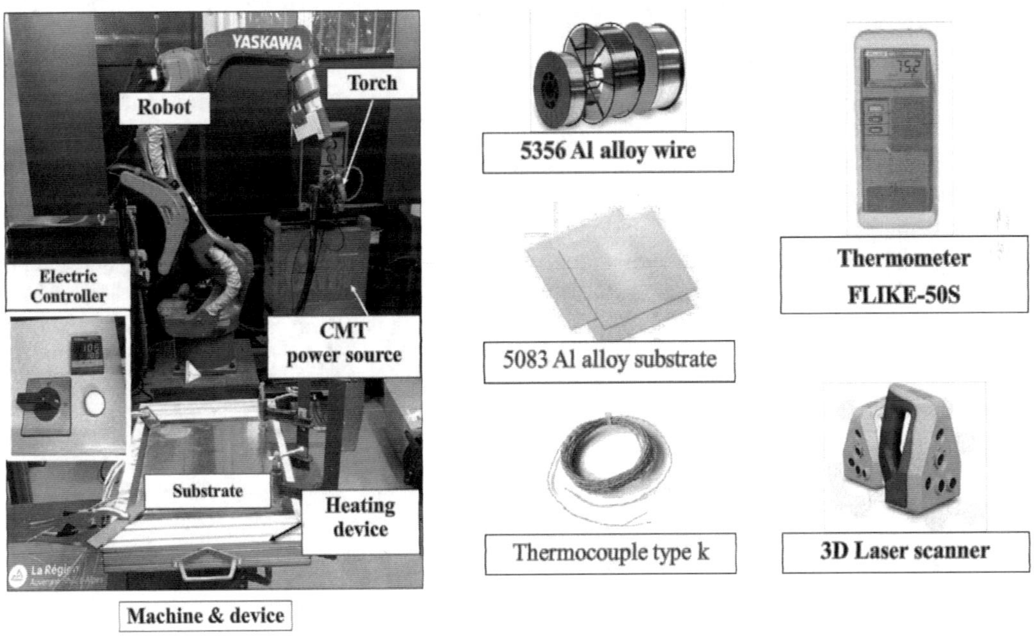

Figure 1. Equipment's and materials for experiments.

2.2. Methodology
In our experiment, welding parameters have been kept constant whereas the starting temperature has been changed, referred as $Tsur$. For the needs of our model, 4 different surface temperatures have been investigated : 100, 200, 300 and 400 °C. Our model is built from geometry evolutions of single bead walls of length 80 mm composed by ten layers, starting from single layer (L1) to tenth layers (L10) totally 10 sample per surface temperature. The example displayed in Figure 2 has been built with a constant starting temperature of the upper surface of 100°C. The strategy chosen is a Zig-Zag deposition

trajectory. The methodology has been set in order to insure that the surface on which a layer is going to be deposited is at the same temperature for each layer. In order to do so, experiment the substrate temperature is first fixed at 100°C and controlled with k type thermocouples. Once it is stabilized, the first layer L1 is deposited. Then, before launching the next layer, a dwell time is implemented in order to let part cooling to occur. The upper surface of this bead is then measured thanks to a thermometer (FLIKE-50S). Three points are controlled: in the middle of the previous layer and at 15 mm from each extremity. Once again, the next layer is deposited after temperature is stabilized. It could thus become necessary to increase the energy input brought by the heating device to reach this temperature of controlled. Indeed, with the increasing distance between the upper surface and substrate, calories are easily dissipated and it could be required to inject more energy to the system. This process is repeated for each layer to insure the exact same temperature of the upper surface before depositing a new layer.

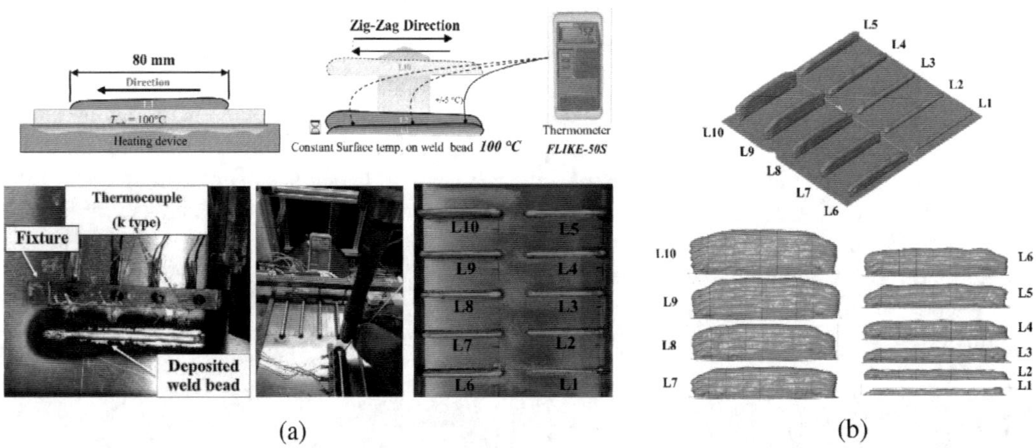

Figure 2. Example : $Tsur$ 100 °C (a) Experimental layout, (b) 3D scan of experiment.

After implementation, our experiments used a 3D laser scanner to verify the dimension of the weld bead in Figure 3, Example: $Tsur$ 100 °C, Cross-section step. The five positions of the cross-section and the distance of 10 mm per cross-section. The result of five cross-sections profile are found that the width (w) and height (h) but still, failing to confirm the weld bead correct dimension. Due to observed that the appearance of the on top weld bead is close to the semi-circle a better approximation. Therefore, As shown in Figure 4, the step circle fitting of L1, L2 to L10. This is mainly due to the use of semi-circle circle fitting with R-squared (R^2) values of the total five cross-sections profile. The obtained were radius (r) and the centre point of the circle (z) leading to finding the weld bead width (w) and the height (h) L1 to L10.

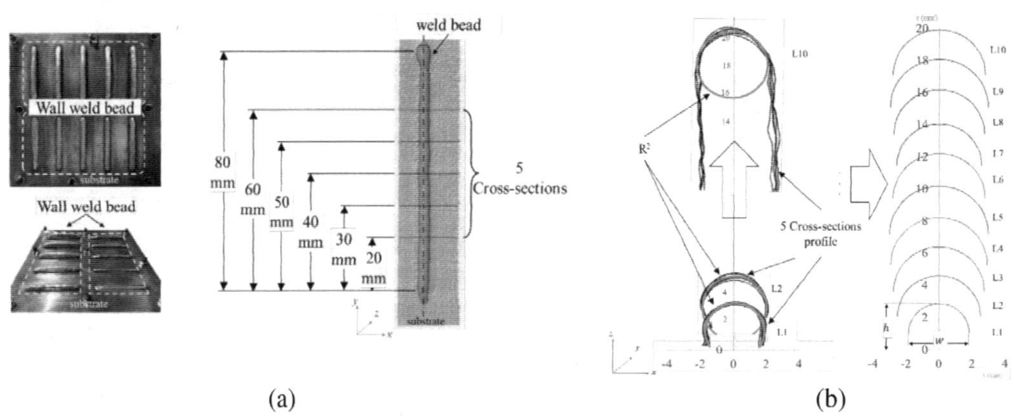

Figure 3. Example : $Tsur$ 100 °C (a) Cross-section position layout, (b) Circle fitting with each layer.

IOP Conf. Series: Materials Science and Engineering 1063 (2021) 012008 doi:10.1088/1757-899X/1063/1/012008

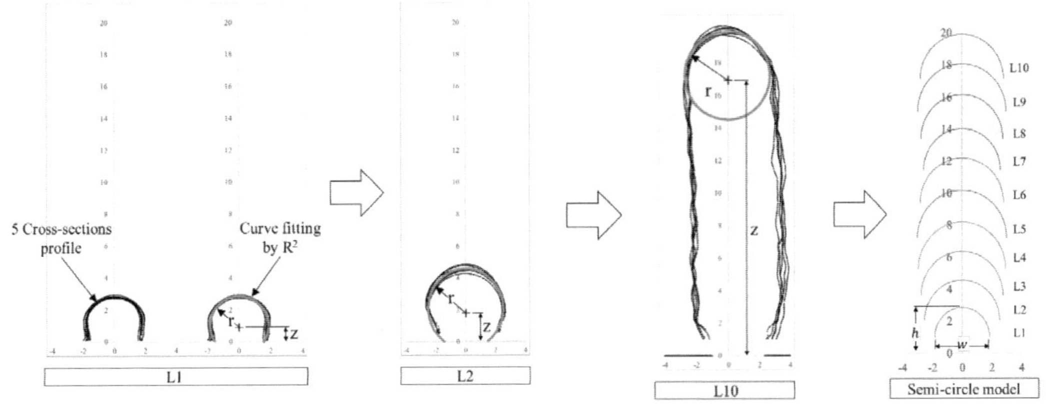

Figure 4. Example : *Tsur* 100°C of L1, L2 to L10, Step circle fitting by R^2 to semi-circle model.

3. Result and Discussion

The weld beads have been built on the semi-circle model shown in Figure 5 *Tsur* 100, 200, 300 and 400 °C. In Table 1. the results were weld bead width (*w*), height (*h*) and average R-squared (R^2).The Avg. R^2 values are observed that close to the dimension of the weld bead. Therefore, it can determined that the semi-circle model can be used in predictions of weld bead. There is still an observation. Especially, the first layer thus temperature increasing this change also induced a variation in *w* and *h* whom even seem linked (as expected): while one is decreasing, the other one is increasing, suggesting that the surface of the cross-section is kept constant. It was found that the first layer has a difference in *w* and *h* compared with another layer due to the flat surface of the substrate. While second to tenth layers, the deposition was started on the curve surface. Thus, the variables that affect to the dimension of weld beads such as surface temperature, the surface beginner of start deposition and welding parameter. The semi-circle model can help the real part dimension of the manufacturing WAAM process.

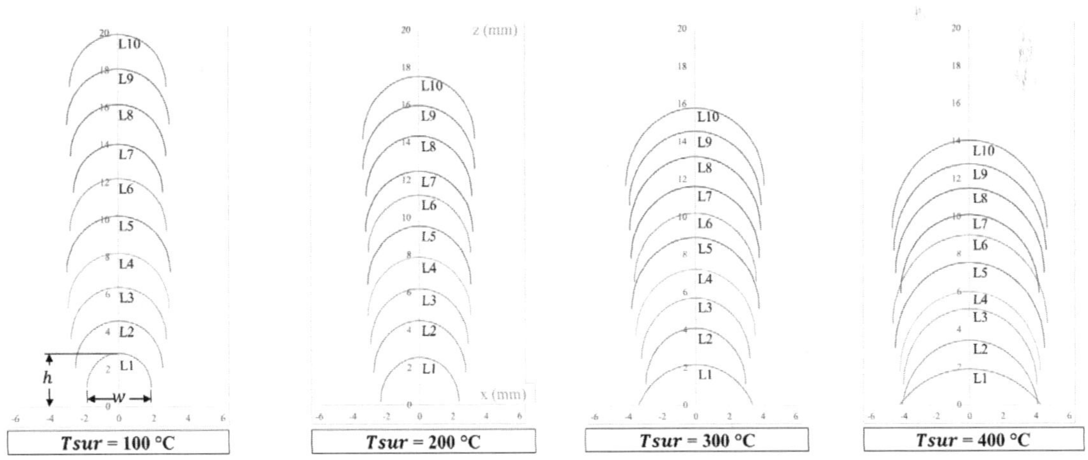

Figure 5. Result of *Tsur* 100, 200, 300 and 400°C.

Table 1. Result of dimensions number of layers compared with control temperature on surface.

Tsur / Number of layers	100 °C			200 °C			300 °C			400 °C		
	Width, *w* (mm)	Height, *h* (mm)	*Avg. R^2 (mm)	Width, *w* (mm)	Height ,*h* (mm)	*Avg. R^2 (mm)	Width, *w* (mm)	Height, *h* (mm)	*Avg. R^2 (mm)	Width, *w* (mm)	Height, *h* (mm)	*Avg.R^2 (mm)
L1	3.06	2.86	0.01	4.68	2.54	0.01	6.78	2.17	0.01	8.54	1.89	0.01
L2	5.02	4.59	0.02	5.52	4.50	0.08	5.92	4.09	0.04	8.36	3.43	0.02
L3	5.50	6.38	0.02	5.84	6.20	0.04	6.40	5.69	0.06	8.06	5.09	0.05
L4	5.84	8.20	0.02	6.16	7.87	0.01	7.06	7.21	0.01	8.50	6.00	0.06
L5	5.98	10.19	0.08	6.18	9.55	0.03	7.54	8.90	0.01	9.02	7.55	0.11
L6	5.58	12.20	0.09	6.14	11.20	0.03	7.22	10.21	0.08	9.34	9.00	0.11
L7	5.14	14.04	0.03	6.44	12.47	0.04	7.60	11.63	0.05	8.36	10.11	0.17
L8	5.50	16.16	0.08	6.34	14.34	0.02	7.76	13.19	0.05	9.00	11.51	0.20
L9	5.92	18.07	0.06	6.70	15.95	0.03	7.80	14.55	0.07	9.20	12.79	0.07
L10	5.58	19.91	0.03	6.70	17.53	0.05	8.20	15.78	0.08	9.36	14.04	0.06

**Tsur* = Surface temperatures
*Avg. R^2 = Average R-squared

4. Conclusions

This paper aims has focused on the development of the semi-circle model apply to weld bead of dimensions for the WAAM process. The relationships have between the different surface temperatures and the dimension of weld beads. Result of the experiments were obtained variables the surface beginner of start on flat surface and another layer on curve surface deposition affect to control of dimension. Thus, it can be helped determine the control dimension of the weld beads. While the welding parameter is constant. This ensures weld bead circle fitting by semi-circle can improve manufacturing efficiency. In future studies, the experiment will be developed to make the weld bead horizontal direction and consider the weld bead geometry with the prediction model varies.

5. References

[1] Donghong D, Zengxi P, Dominic C and Huijun L 2015 A multi-bead overlapping model for robotic wire and arc additive manufacturing (WAAM) *Robotics and Computer Integrated Manufacturing* **31** pp.101-110

[2] Donghong D, Zengxi P, Dominic C and Huijun L 2015 Wire-feed additive manufacturing of metal components: technologies, developments and future interests *The International Journal of Advanced Manufacturing Technologyl* **81**(1-4) pp.465-481

[3] Donghong D, Zengxi P, Dominic C, Huijun L, Stephen V D and Nathan L 2016 Bead modelling and implementation of adaptive MAT path in wire and arc additive manufacturing *Robotics and Computer-Integrated Manufacturing* **39** pp.32-42

[4] Kisarev AV and Kobernik N V 2020 Study on formation of aluminum alloy thin wall produced with WAAM method under various thermal conditions *IOP Conference Series: Materials Science and Engineering* **759** p 012014

[5] Bintao W, Donghong D, Zengxi P, Dominic C, Huijun L, Jian H and Zhenyu F 2017 Effects of heat accumulation on the arc characteristics and metal transfer behavior in Wire Arc Additive Manufacturing of Ti6Al4V *Journal of Materials Processing Technology* **250** pp.304-312

[6] Dongqing Y, Gang W and Guangjun Z 2017 Thermal analysis for single-pass multi-layer GMAW based additive manufacturing using infrared thermography *Journal of Materials Processing Technology* **244** pp.215-224

[7] Selvi S, Vushvaksenan A and Rajasekar E 2018 Cold metal transfer (CMT) technology - An overview *Defence Technology* **14**(1) pp.28-44

[8] Cong B, Ouyang R, Qi B and Ding J 2016 Influence of Cold Metal Transfer Process and Its Heat Input on Weld Bead Geometry and Porosity of Aluminum-Copper Alloy Welds *Rare Metal Materials and Engineering* **45**(3) pp.606-611

IOP Conf. Series: Materials Science and Engineering 1063 (2021) 012008 doi:10.1088/1757-899X/1063/1/012008

[9] Cong B, Ding J and Stewart W 2015 Effect of arc mode in cold metal transfer process on porosity of additively manufactured Al-6.3%Cu alloy *The International Journal of Advanced Manufacturing Technology* **76**(9-12) pp.1593-1606

IOP Conf. Series: Materials Science and Engineering 1063 (2021) 012009 doi:10.1088/1757-899X/1063/1/012009

Applying value stream mapping in packaging industry: a case study

O Joochim[1,*] and S Jungthawan[2]

[1]Institute of Field Robotics, King Mongkut's University of Technology Thonburi, Bangkok, Thailand
[2]Department of Civil Engineering, Faculty of Engineering, King Mongkut's Institute of Technology Ladkrabang, Bangkok, Thailand

*E-mail: orapadee.joo@kmutt.ac.th

Abstract. The aim of this paper is to increase the productivity of a packaging manufacturing process by using lean manufacturing technique. The current system of this packaging manufacturing process has been mapped using Value Stream Mapping (VSM) tool. This tool is a lean manufacturing tool, and the wastes in the current system are identified. The case study shows how the VSM tool has been utilized for identifying and decreasing the waste in the manufacturing process resulting of the total cost saving of 3,247,251 Thai Baht (THB) per year.

1. Introduction

These Thai packaging industry has played a significantly important role in the economic and industrial sector of Thailand due to the disruption of traditional culture in the sense of a way of life and the extremely fast pace of urban living increased demand for a greater level of convenience. The packaging industry in Thailand is approximated to expand from 61.4 billion units in 2018 to 71.3 billion units by 2023. This expansion can be mainly attributed to the expansion of the country's economy, increasing disposable incomes, and spending capabilities, leading to the increased consumption of packaged products. In order to enhance the competitiveness, therefore, there is the need for the packaging companies to look at best practices to improve their performance [1].

Value Stream Mapping (VSM) is an improvement tool used for identifying, illustrating and reducing wastes, as well as creating flow in the manufacturing process. As mentioned by Rother and Shook [2], the companies must think about the flow, rather than implementing thoughtful manufacturing systems or separated enhancement processes. The VSM approach has declared itself effective by concentrating its considering flow. VSM is a framework to visibly represent the material flow and information through the manufacturing process. The aim is to determine and increase value addition together with removing or decreasing non-value added activities. VSM is defined as a part of the lean manufacturing tools. The most efficient VSM are mapped by a cross-functional team that has people from various activities in the value stream being assessed. Moreover, the VSM development is a part of the training applied for deeper understanding of the application of lean manufacturing implementation [3,4,5,6].

This research is a case study of VSM implementation at one of the biggest packaging manufacturers in Thailand. The paper intensely investigates the case study company's manufacturing processes for identifying waste streams. A current state map is developed to identify how materials and information flow. Although, this research has been limited to a packaging manufacturer, obtainable literature demonstrates that VSM can additionally be applied to all other sectors with essential adjustments [7].

It will yield results as organizations are seeking ways for increasing the value of their products and leading to better and happier customers by eliminating wastes.

2. Case Study Manufacturer

The case study company is the manufacturer and distributor of PVC hard plastic sheets. With a production capacity of 18,000 tons per year, the manufacturing process starts from preparing raw materials, mixing ingredients according to the recipe, making the plastic extrusion with the calender machine and cutting the plastics to size according to the customer needs, and finally packaged them for delivery (See Figure 1).

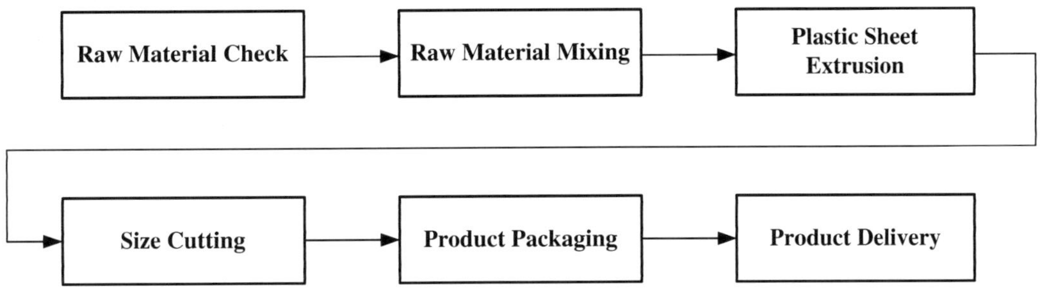

Figure 1. Manufacturing Process.

The products of the manufacturer are rigid plastic sheet products for various processing applications. They can be divided into 4 main categories: 1) plastic cards such as credit cards, ATM cards, food grade and medical plastic infection molding, 2) general printing such as packaging and plastics for upholstered furniture. The manufacturer has both domestic and international customers including Asia, USA and Europe. The Company's suppliers of raw materials come from various sources in Thailand and other countries.

The products of the manufacturer are preferred by both domestic and international customers. However, with the current competition conditions, the manufacturer deals with the problems. Both domestic and foreign competitors reduce the prices of their product. The competitors import the products from abroad to sell at lower prices. Moreover, the sales forecast from the sales department does not have complete product details resulting in purchase errors leading to inventory wastes for both raw materials and finished products. The Product delivery from raw material suppliers has also uncertain lead time and unstable control in the production process causing a delay in the delivery of work.

The manufacturing is a semi-continuous production. (semi-continuous process) and the large batch production. There are many processes that the time to set up the machine is quite high, more than 1 hour, affecting the flexibility of production (operational flexibility). The work study is used to analyse the production process. For the warehouse, advanced warehouse management concepts are used, i.e., mathematical modeling, order quantity determination, order lead time and reorder point to manage space and inventory in the warehouse.

The land transportation of goods by roads and water ways at Laem Chabang Sea Port is used with the internal organization communication using the information system. However, there is missing information in some departments and the lack of integrated supply chain management within the organization.

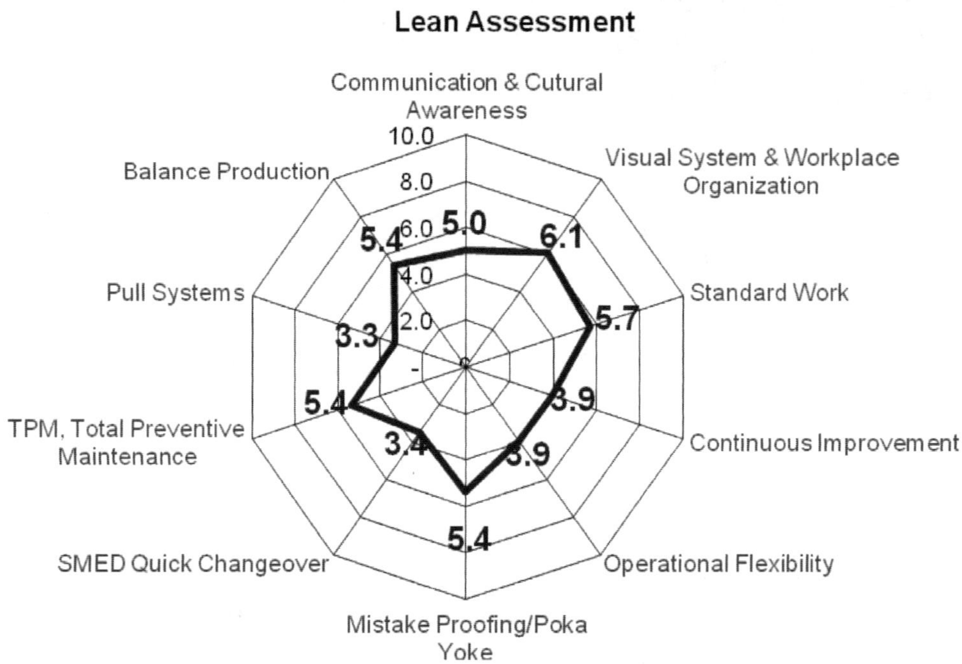

Figure 2. Lean Assessment before the Improvement.

Results of the initial assessment of the manufacturer in accordance with lean Manufacturing guidelines compared with lean best practices, pointing out the areas that can be further improved. For example pull systems, due to the nature of the production machinery is a semi-continuous process. For Single Minute Exchange of Die (SMED), there are several processes that have a very high set-up time of more than 1 hour. The production flexibility (operational flexibility) as well as the continuous improvement are additionally should be improved, as shown in Figure 2.

3. Developing Value Steam Mapping

Value Stream Mapping (VSM) is a tool based on the fundamental principle of Lean Production or Lean Enterprise. This tool helps organizations for decreasing lead times, inventory, enhancing quality and get better on-time deliveries and utilization resources. Thus, reduction of non-productive activities (viz., wastes) keeps resources as well as accepts resource reallocation in order to improve throughput and profitability [6]. A concept of VSM is illustrated in Figure 3. Current-state map (CSM) is developed to identify how materials and information flow from the start to the endpoint of the process [5]. The detailed explanation of VSM can be investigated from the book of K. Martin and M. Osterling [8].

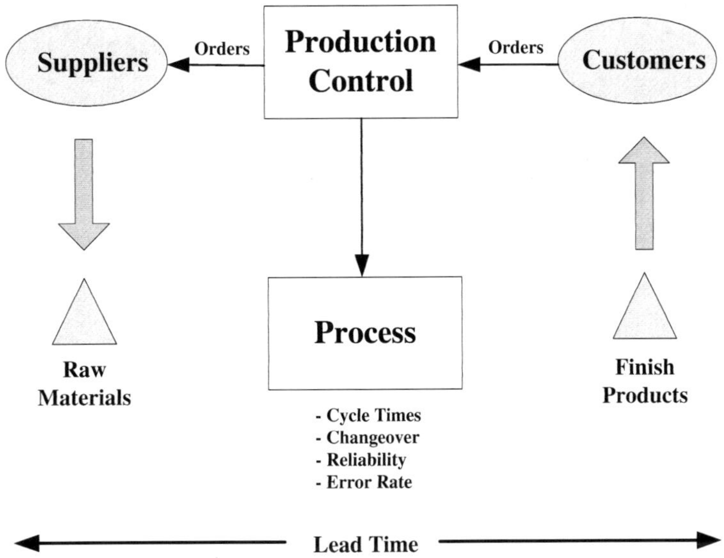

Figure 3. Value Stream Map Concept [5].

CSM of Packaging manufacturing process of a case manufacturer is shown in Figure 4. The line is a dedicated flow line, and several operations needed for manufacturing packaging products are also illustrated in Figure 4. CSM is the baseline perspective of the existing situation from which all improvements are determined. The map shows the product movement from suppliers through customers. At the lowest point of CSM, the timeline demonstrates the Value-Added (VA) and Non-Value Added (NVA) Process Times from the entry point of manufacturing (raw materials) into the line to the finished products. The results of this study comprise of the production flow, the waste pointed out and listed CSM. Due to prioritizing waste reduction targets, the analysis of CSM in this research can be grouped into three categories, which are raw material mixing process, calender machine adjustment and finished product inspection and movement processes. Table 1 shows the summary of the problems of the manufacturing process for improvement.

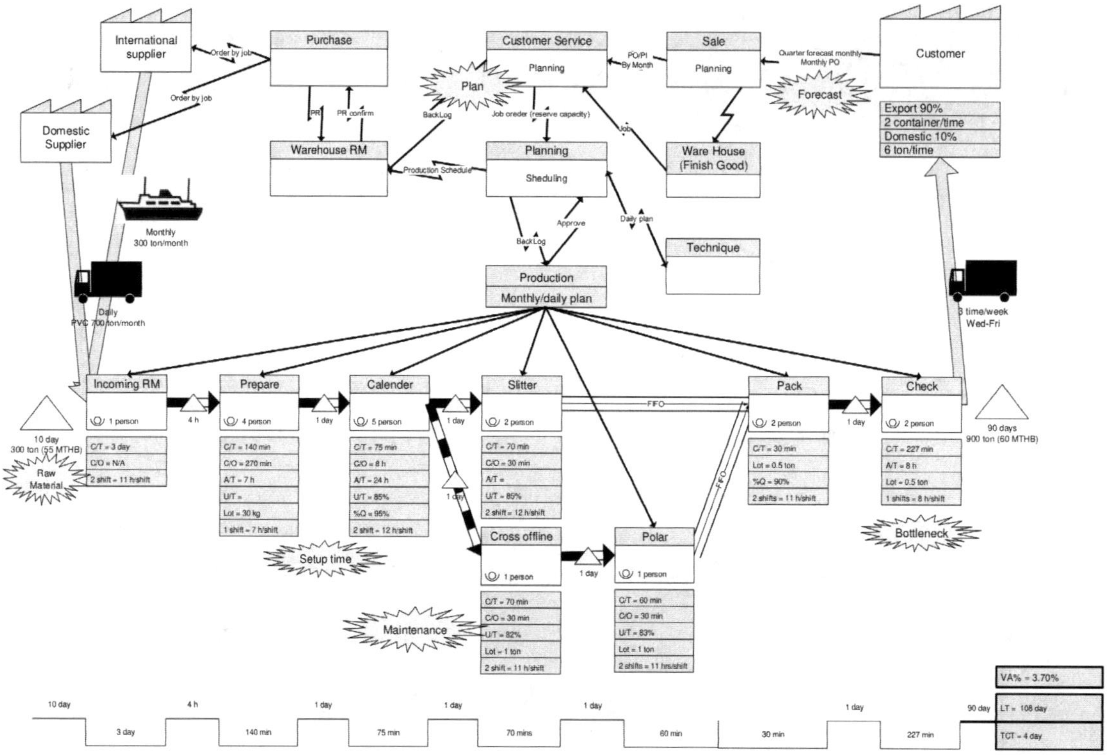

Figure 4. CSM of the Case Study Manufacturer.

Table 1. The Problems Investigated from VSM for Improvement.

Improvement Activities	Lean Tools	Baseline	Target	Result	Diff	Cost Saving	
						Hard Saving	Potential Saving
1. Raw Material Mixing Process	Work Study	140 Minutes	6% Decreased	12.66% Decreased	6.66%	17,935 baht/year	-
2. Calender Machine Adjustment	Faster Setting on the Machine	8 Hours	10% Decreased	24.5% Decreased	14.5%	3,216,000 baht/year	-
3. Finished Product Inspection and Movement Processes	Work Study	227 Minutes	10% Decreased	11.45% Decreased	1.45%	13,316 baht/year	-
					Total	3,247,251 baht/year	-

4. Conclusion

After the process improvement, the case study manufacturer can enhance competitiveness. It can be concluded that VSM is a valuable planning tool and can be utilized to develop and implement lean improvement of packaging manufacturing processes. Figure 5 shows the improvement results. By optimizing production to aim for lean production standards, raw material mixing process is improved by 10% faster for raw material mixing process, calibration methods for the calender machine with 25% reduced set-up times, and 12% faster for the inspection and movement of finished goods. The total cost savings is 3,247,251 baht / year.

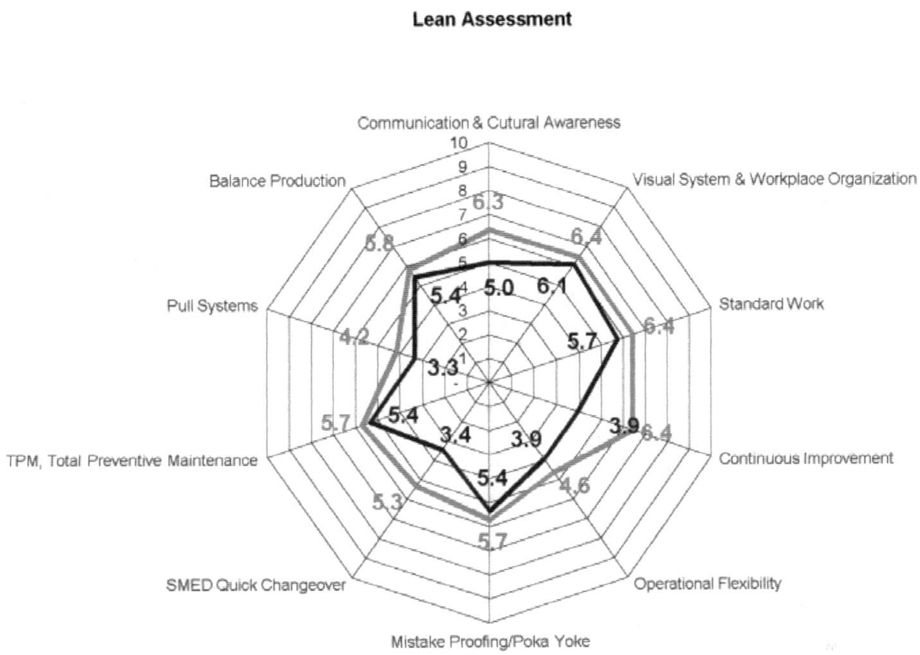

Figure 5. Lean Assessment after the Improvement.

References

[1] GlobalData 2019 *Trends and Opportunities in the Thai Packaging Industry: Analysis of Changing Packaging Trends in the Food*, Cosmetics and Toiletries, Beverages and Other Industries

[2] M. Ben Rabha, M.F. Boujmil, M. Saadoun, B. Bessaïs, Eur. Phys. J. Appl. Phys. (to be published)

[3] G. H. Martins and M.G. Cleto 2013 *Value Stream Mapping and Earned Value Analysis: A Case Study* the Paper Packaging Industry in Brazil (22nd International Conference on Production Research, Brazil)

[4] W.M. Goriwondo S. Mhlanga and A. Marecha 2011 *Use of the Value Stream Mapping Tool for Waste Reduction in Manufacturing. Case Study for Bread Manufacturing in Zimbabwe* (Proceedings of the 2011 International Conference on Industrial Engineering and Operations Management, Kuala Lumpur, Malaysia)

[5] F.C. Garcia *Using Value Stream Mapping as a Strategic Planning and Implementation Tool* (White Paper, Business Solution & Engineering Services, Advent Design Corporation, USA)

[6] O. Joochim and S. Jungthawan 2015 *An Application of Value Stream Mapping in Ceramic Tile Industry: A Case Study* (Applied Mechanics and Materials)

[7] B. Singh, S.k. Garg and S.k. Sharma 2010 *Value Stream Mapping: Literature Review and Implications for Indian Industry* (Int J Adv Manuf Technol)

[8] K. Martin and M. Osterling 2014 *Value Stream Mapping: How to Visualize Work and Align Leadership for Organizational Transformation* (McGraw-Hill Education Books)

IOP Conf. Series: Materials Science and Engineering 1063 (2021) 012010 doi:10.1088/1757-899X/1063/1/012010

The Industry's Perspective of Suitable Tasks for Human-Robot Collaboration in Assembly Manufacturing

Patrik Gustavsson and Anna Syberfeldt

University of Skövde, Högskolevägen, Box 408, 541 28, Skövde, Sweden

patrik.gustavsson@his.se

Abstract. Human-robot collaboration (HRC) is the concept of combining a human and a robot into the same production cell and utilize the benefits of both. This concept has existed for more than a decade, but there are still quite few implementations of HRC within the manufacturing industry. One reason for this is the lack of knowledge when it comes to suitable tasks for HRC. Current research studies on the topic are mainly based on theoretical reasoning and/or research experiments, and little is known about what the industry perceive as suitable tasks for HRC. Therefore, this paper aims to investigate this and find out what industrial actors thinks are the most value-adding tasks for a human and a robot to carry out together. An in-depth interview study is undertaken with two companies and shop-floor operators, production engineers and automation engineers are interviewed. The result of the study pinpoints a number of tasks that the companies thinks are beneficial for HRC, which can serve as a guideline for other manufacturing companies considering to implement HRC.

1. Introduction

Human-robot collaboration combines the strengths of both humans and robots into a hybrid production cell [1]. There are several application areas in which HRC are advantageous, of which assembly manufacturing is one of the major [2]. In current implementations of HRC there are, however, often a limited interaction between the human and the robot in order to ensure the safety of the human [3]. This is a drawback since to fully utilize the potential of HRC, the human and the robot should interact and not only work side-by-side or in sequence with each other. To realize work cells in which the human and the robot are truly working together with each other it must be known what tasks that are suitable for such collaboration. There are a number of research papers that presents studies on the topic, but these are mainly based on theoretical reasoning and/or research experiments that investigate the suitability of various tasks. There is not much knowledge about what manufacturing companies perceive as suitable tasks for HRC, and this paper therefore aims to address this question. The paper is specifically focused on assembly manufacturing as this is a main application area for HRC. An in-depth interview study is undertaken with one manufacturing company and one automation integrator company. In these companies, shop-floor operators, production engineers and automation engineers are asked to give their view on what they perceive as suitable tasks for HRC. The manufacturing company is a large company producing quick connect couplings having production sites in several different countries around the world. The study takes place in one of their sites where these couplings are assembled. The automation integrator company produces automation solutions for manufacturing industries all over Sweden. At the site of focus, automation and robotic solutions are designed and implemented.

In the next chapter, more information about HRC is given for the reader not yet familiar with the topic. The paper then continues by describing the set-up for the interview study in chapter 3 and 4. The

IOP Conf. Series: Materials Science and Engineering 1063 (2021) 012010 doi:10.1088/1757-899X/1063/1/012010

results from the interviews are presented and analyzed in chapter 5. Chapter 6, finally, summarized the conclusions from the study.

2. Background

Research has been done in the area of HRC for many years and involves multiple disciplines such as interaction, safety, path-planning and task allocation to mention but a few. This chapter gives the basic knowledge of HRC necessary to understand the paper, and also describes examples of research studies in which HRC workstations have been implemented.

In HRC a human and a robot share workspace to complete a task, and the idea is to combine the benefit humans and robots into one workstation. Since traditional industrial robots pose a danger to a human when they share the same workspace, industrial robot manufacturers have started to produce collaborative robots to overcome this safety issue. Collaborative robots are robots that can be used in collaborative operations as defined in the technical specification ISO/TS 15066 [4], created by the International Organization for Standardization. Collaborative robots are generally lightweight industrial robots that include force limitations on all joints to make them suitable for HRC. In the literature, there are plenty of studies on HRC and some of the most interesting of these are presented in the following of this chapter.

Sadrfaridpour and Wang [1] presents a framework for human-robot interaction and showcase this framework in a workstation where a human and a robot collaborates. The task that the human and robot executes is to assemble three parts. The human fetches one part and places it in the workspace, the robot then fetches two other parts and places them onto the first part, and finally the human screws the parts together.

Hietanen et al. [5] presents another study in which augmented reality-based interaction is used for collaborating with a robot. The authors propose an interactive user interface for displaying digital information that is projected onto the real world. A workstation is implemented in which part of an engine is to be assembled, adopted from a real-world case. The task consists of five sub-tasks, of which three are manual, one is handled by the robot and one requires the collaboration of both human and robot. In the collaborative sub-task the robot brings a component and activates hand-guidance, which allows the human to guide the robot by hand to position the component.

Peternel et al. [6] presents and demonstrates a multi-modal robot teaching framework which is used to train a robot to physically perform cross-cut sawing motions with a companion human. Their work is an interesting example of when a human and a robot physically interacts to execute a task. In the task, the human and the robot are on opposite sides of the saw only exerting force in a dragging motion. While the robot is dragging the saw, the human mainly follows, and while the human is dragging the saw, the robot mainly follows.

De Gea Fernández et al. [7] presents a multimodal whole-body tracking system used for gesture recognition, intention recognition and collision avoidance. These technologies are implemented on a HRC workstation combining two collaborative robot arms where. In the task, adopted from a real-world case, the human signals the robotic system with gestures to either start or pause the process, or activating a collaboration mode. When the robot cell starts, the two robots assembles the parts while ensuring the safety for the operator by tracking the human and actively avoiding collision. If an inspection is needed, the operator activates the collaborative mode in which the robot picks up the last assembled product and positions it in a comfortable position for the operator to inspect it.

The summarized papers are some examples of studies that show the great potential of HRC in different application areas, and that the technological advancement in robotic platforms have made it possible to implement HRC within the manufacturing industry [2,8]. So far, the identified tasks within HRC are, however, mainly based on theoretical reasoning and/or research experiments, and as previously discussed more investigation is needed to further identify tasks that will be beneficial for HRC.

3. Interview subjects

When it comes to relevant interview subjects for the study there are two types of persons that are of interest: persons that are working with assembly tasks that can potentially be made collaborative, and persons that are working with the planning and/or construction of HRC cells. As previously mentioned, the study is focused on assembly tasks specifically. When identifying interview subjects, three categories of professions were identified:

Automation engineer: Persons with this profession can provide knowledge of the automation process, what can and cannot be constructed, and the capabilities of robots including collaborative robots. In the study, automation engineers with knowledge of automation and development of collaborative solutions were sought after.

Production engineer: Persons with this profession have an overall knowledge and responsibility for the production line. They can therefore provide insight into the production and technical aspects of assembly and what can be improved with the help of an assistant, such as a collaborative robot. Production engineers know the process, the history and development over time in the factory, which gives them a different perspective to that of the automation engineer.

Operators working with assembly: Persons with this profession are relevant because their daily work consists of assembly and they are therefore experts on the process as well as its shortcomings and possibilities. They have hands-on experience of the process beyond that of automation and production engineers and should be able to provide good insight into the current process and its opportunities for improvement. Since they work with assembly daily, they can also provide information on what parts of the assembly process that feel inefficient, stressful, or present ergonomic issues.

In total there were ten participants: four automation engineers, three production engineers and three operators. These participants are further on referred to as A1-A4 for the automation engineers, P1-P3 for the production engineers and O1-O3 for the operators.

4. Interview structure

The interview study was conducted according to the procedure in Figure 1. The study used a semi-structured interview approach, with core questions focused on identifying tasks suitable for HRC. A semi-structured interview approach was selected because it gives the participant more freedom to discuss the topic as they want, but still restricts them to the main theme if they stray too far [9]. The questions were open-ended to avoid guiding the thought processes of the participants. The questions were also adapted to the occupational role of the participant.

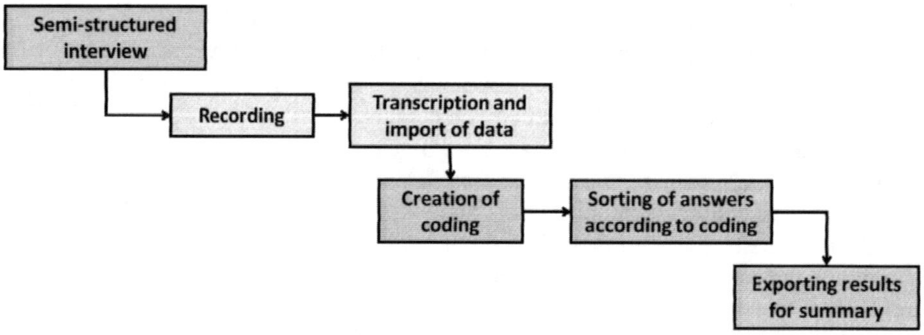

Figure 1. The process of collecting and analysing data in the interview study to extract knowledge of tasks suitable for HRC.

It is well known that the language used during an interview is important, and the interviewer must be able to adapt the language to the interview subject without mimicking or imitating people in any way [10]. The interviewer should thus be able to adapt in such a way that the question and the message become clear without making the situation unnatural. Having this in mind, the interview questions were adapted depending on the occupational role (see section 3) of the interview subject.

4.1. Interview process

All interviews started with an introduction presenting the interviewers, the purpose of the interview, obtaining permission to record, and assuring confidentiality. After getting a signature from the participant, the interview was recorded using a smartphone for the purpose of later transcribing and analyzing the interviews, see figure 1. Then the semi-structured interview started following an interview form which was structured with a set of questions to simplify the process for the interviewer. As the last part of the interview the interview subjects were asked if they had any comments, questions or anything else to add.

Three interview forms were created for this study based on the identified professions in chapter 3, these were used for the interviewer to structure the questions. Each form consisted of a set of phases and each phase had a set of questions with alternative follow-up questions depending on the answers. All forms consisted of the following four phases, with the exception that phase 2 was not used for automation engineers:

Introduction: In the introduction phase, the interview subjects were asked questions with simple answers. This was to warm them up and make them more comfortable in answering the more complex questions that were following [9,11]. The questions used in this phase were the following: "What is your occupation?", "How long have you worked in your current occupation?", "How long have you worked in this company?" and "What was your previous occupation?".

Current status: In this phase, questions were asked about existing problems within the assembly manufacturing line. As the automation engineers were not connected to a specific line or station, this phase was excluded for them. The questions in this phase were used to get an understanding of the interview subject's view of the current situation. The answers were expected to pinpoint specific tasks that are perceived as time consuming, requiring high precision, working with small or heavy objects, working with parts that are difficult to handle parts, ergonomically stressful tasks, or other similar complexities. The questions in this phase are also asked to provide an image of the interview subject's work and to prepare the interview subject to think about tasks that may be relevant to the core question.

Core questions: This was the main phase of the interview, where the purpose was to gain knowledge about what tasks that are suitable for HRC. The core question needed to be carefully formulated for the production engineers and operators since it could not be expected that these were familiar with the possibilities of collaborative robot and HRC. Therefore, the question used for these two categories were formulated as "What could a colleague or an extra person assist you with to facilitate or simplify your task?".

The questions for the automation engineers directly asked about identifying possible HRC tasks, as this profession were known to have prior knowledge about HRC. They were first asked whether they had previously practically worked with implementing HRC or collaborative robots. If they had, they were asked for more information about that implementation, otherwise they were asked in which tasks in assembly manufacturing they could see possibilities using HRC.

Future analysis: In this phase, questions were asked on how the interview subjects thought that the future would look like. For production engineers and operators, questions were asked about their perception of HRC. If they had no idea what HRC could look like, the interviewers gave a brief explanation. For automation engineers, questions were asked on how they would want to work with collaborative robots and HRC, how it could be tested virtually, and how the robot needs to be further developed to become a more common solution in industry.

4.2. Analysis of interview

After the interviews had been conducted all recordings were transcribed, which simplifies the process to search through and analyze the data [9]. The transcribed text was imported into Dedoose, an online app used for analyzing qualitative and mixed methods research with text, photos, and audio. Codes were created to categorize the answers of each interview subject. The information extracted from the interviews were then carefully analyzed.

From the transcribed material, root codes were created to categorize the answers into "experience", "tasks" and "improvements". Sub codes were then created to further divide the code "tasks" into the type of problem that is correlated with the described task. The reason for this code creation is to identify the type of problem within tasks that HRC can be used to alleviate.

5. Results

Based on the results from the analysis, six main categories of tasks were identified where support for the operator would be suitable. These categories are listed in table 1 in alphabetical order. The table shows the opinions of the interview subjects listed in each column, where a checkmark is added for each category of tasks that they mentioned can benefit from HRC.

Table 1. The interview subjects' opinions on potential tasks that can benefit from HRC divided into six categories. A1-A4 are automation engineers, P1-P3 production engineers and O1-O3 operators.

	A1	A2	A3	A4	P1	P2	P3	O1	O2	O3
Difficult to operate	✓				✓		✓			
Logistic inefficient			✓	✓	✓	✓	✓	✓	✓	✓
Non-ergonomic	✓	✓	✓	✓	✓	✓	✓	✓	✓	✓
Product variation			✓					✓		✓
Time consuming		✓	✓		✓				✓	
Uneven quality	✓	✓	✓			✓				

From the results in Table 1, all ten participants discussed non-ergonomic tasks and eight of the participants discussed logistic inefficient tasks. For each of the remaining categories, between three to four participants discussed tasks belonging to respective category.

In the following subsections each of the categories are described in more detail with examples of tasks that the interview subjects discussed.

5.1. Difficult to operate

Automation engineer 1 mentioned that in tasks that requires skill and dexterity with the fingers, a collaborative robot could assist the operator – especially if the task is difficult to perform using another machine or tool. Operator 2 also mentioned the same aspect and gave the entering of small screws as example. Two out of the three production engineers also mentioned tasks where it is difficult to reach as possible tasks when a collaborative robot could be beneficial.

5.2. Logistic inefficient

Most participants mentioned that a collaborative robot could assist in tasks related to logistics around the assembly stations. All operators discussed material preparation and preparation for the next product as an example, and that assistance with surrounding equipment could facilitate such work. Operator 3 mentioned that assistance in loading material and preparing for the next product could reduce downtime and improve productivity. Production engineer 2 mentioned tasks that involve placing details in the station for identification. Production engineer 1 suggested that a collaborative robot could function as a third hand, even in logistics. Automation engineers 3 and 4 talked about a HRC cell that had previously been built and that they used a mobile solution to drive materials to the station. This too could be in the form of an intelligent collaborative robot in the future.

5.3. Non-ergonomic

All participants mentioned that ergonomic relief for the human is a good potential utilization of a collaborative robot, as seen in table 1. One specific task that they discussed was the non-ergonomic entering of screws. Production engineer 1 mentioned that the entering of screws is tiring for the wrists and has affected the employees negatively from an ergonomic perspective. Automation engineer 2 and 3 also mentioned similar entering tasks in their assembly lines. A recurring example of tasks that were mentioned are those that involve repetitive heavy lifting. Automation engineer 4, who had also

IOP Conf. Series: Materials Science and Engineering 1063 (2021) 012010 doi:10.1088/1757-899X/1063/1/012010

participated in a project which involved a HRC cell, said that there is a task at their station where the collaborative robot and the human together lift a heavy metal piece. Operator 3 also mentioned that possible tasks for a collaborative robot include when the operator is forced to work at an uncomfortable height.

5.4. Product variation
During the interviews, all automation and production engineers talked about how future production will require increased flexibility. One of the reasons for this was the increased customer demand for product variation in the market. Automation engineer 3 said that the vision for collaborative solutions should be that they are able to cope with the preparation process even if the products change. The robot should be intelligent enough to be updated automatically and be aware of the product type it should adapt to. Automation engineer 3 explained that in future production where more robots and complex systems will be present, it will be difficult for humans to have all the information in their head. Therefore, it would be advantageous if the robot could take more responsibility regarding product changes and preparation for new details. Operator 1 and 3 did not mention any specific tasks but said that smooth assistance in complicated switches between products is something that would facilitate the work.

5.5. Time consuming tasks
Regarding tasks that are time-consuming for the operator, both automation engineer 2 and 3 mentioned parts where many screws must be tightened. Automation engineer 2 thought that such a task could just as easily be done by a robot. Automation engineer 3, who participated in the project with the HRC cell, said that one of the tasks that the collaborative robot performs at their test station is tightening 24 screws. Both operator 2 and production engineer 1 stated that a collaborative robot could contribute to efficiency in the tasks at the stations as it involves several parts during assembly.

5.6. Uneven quality
Regarding quality, automation engineer 2 mentioned a collaborative solution that their company provided which glued dashboards. This was a task previously performed manually, but to reach a uniform product quality, a collaborative robot was implemented. Production engineer 2 also mentioned an existing solution in the company's production. A YUMI robot checks the quality of the product as one of the station's tasks before the product continues in the production line. Neither automation engineer 1 nor automation engineer 3 mentioned any specific task. However, automation engineer 1 mentioned that in general automation is used to achieve better product quality. Automation engineer 3 also mentioned improvement in product quality.

6. Conclusion
This paper presents an in-depth interview study that investigated what industrial actors think are the most value-adding tasks in HRC. Research on HRC has been active for many years but despite this, there are still little knowledge of what tasks are suitable for HRC and the results from this interview study can serve as guideline for other manufacturing companies that consider implementing HRC. Two companies participated where three shop-floor operators, three production engineers and four automation engineers were interviewed. From these interviews, six categories of tasks were identified as being beneficial for the human and robot to carry out together. These categories are tasks that are difficult to operate, logistic inefficient, non-ergonomic or time consuming and tasks that have product variation or uneven quality. Two of the categories of tasks that could benefit from HRC were discussed more than the others, all ten participants discussed about non-ergonomic tasks and eight participants discussed about logistic inefficient tasks. For each of the remainder of categories between three to four participants discussed tasks belonging to respective category.

Acknowledgment
We would like to give a thank you to the participating companies and their personnel that made this study possible. A special thanks goes to Lisa Lexe and Rebecca Nilsson, who prepared and executed this interview study.

References
[1] Sadrfaridpour B and Wang Y 2017 Collaborative Assembly in Hybrid Manufacturing Cells: An Integrated Framework for Human-Robot Interaction *IEEE Trans. Autom. Sci. Eng.* **PP** 1–15
[2] Villani V, Pini F, Leali F and Secchi C 2018 Survey on human–robot collaboration in industrial settings: Safety, intuitive interfaces and applications *Mechatronics* **55** 248–66
[3] Fast-Berglund Å, Palmkvist F, Nyqvist P, Ekered S and Åkerman M 2016 Evaluating Cobots for Final Assembly *Procedia CIRP* **44** 175–80
[4] ISO 2016 *ISO/TS 15066: Robots and robotic devices — Collaborative robots* (International Organization for Standardization)
[5] Hietanen A, Pieters R, Lanz M, Latokartano J and Kämäräinen J-K 2020 AR-based interaction for human-robot collaborative manufacturing *Robot. Comput.-Integr. Manuf.* **63** 101891
[6] Peternel L, Petrič T, Oztop E and Babič J 2014 Teaching robots to cooperate with humans in dynamic manipulation tasks based on multi-modal human-in-the-loop approach *Auton. Robots* **36** 123–36
[7] de Gea Fernández J, Mronga D, Günther M, Knobloch T, Wirkus M, Schröer M, Trampler M, Stiene S, Kirchner E, Bargsten V, Bänziger T, Teiwes J, Krüger T and Kirchner F 2017 Multimodal sensor-based whole-body control for human–robot collaboration in industrial settings *Robot. Auton. Syst.* **94** 102–19
[8] Ajoudani A, Zanchettin A M, Ivaldi S, Albu-Schäffer A, Kosuge K and Khatib O 2018 Progress and prospects of the human–robot collaboration *Auton. Robots* **42** 957–75
[9] Oates B J 2005 *Researching information systems and computing* (Sage)
[10] Trost J 2010 *Kvalitativa intervjuer* (Lund: Studentlitteratur)
[11] Kylén J-A 2004 *Att få svar: intervju, enkät, observation* (Stockholm: Bonnier utbildning)

Comparison of Void Fraction on Annular and Churn Two-Phase Flow Patterns in Mini Pipe with a Slope of 15° Against Horizontal Position

Sukamta[1,*], Sudarja[1], Sri Sundari[2]
[1]Department of Mechanical Engineering, Faculty of Engineering, Universitas Muhammadiyah Yogyakarta, Jl. Brawijaya, Tamantirto, Kasihan, Bantul, Daerah Istimewa Yogyakarta, Indonesia, 55183
[2]Department of Medicine, Faculty of Medicine and Health Science, Universitas Muhammadiyah Yogyakarta, Jl. Brawijaya, Tamantirto, Kasihan, Bantul, Daerah Istimewa Yogyakarta, Indonesia, 55183

[*]E-mail: sukamta@umy.ac.id

Abstract. The two-phase flow is part of a multiphase flow, where the flow can consist of gas/vapor and liquid phases, which flow simultaneously. This study aims to compare the characteristics of void fraction values in the annular and churn flow patterns that can be known from the parameters of the gas and liquid superficial velocity, the length of the flow pattern, and the frequency of the flow pattern. But, in this study, the focus is on the superficial velocity of both gas and liquid. Tests were carried out on mini pipes with a diameter of 1.6 mm and a length of 160 mm with a slope of 15° to the horizontal position. Working fluid used was air and water mixed with glycerin with a concentration of 40%, 50%, 60%, 70%. Analysis of research results using the probability distribution function analysis with the help of Matlab and Microsoft Excel programs. The results showed that the comparison of the value of the void fraction of the annular and churn flow patterns was significantly influenced by several factors, namely the superficial velocity of the gas, the superficial velocity of the liquid, and the viscosity of the liquid. The higher the superficial velocity of the gas, the higher the value of the void fraction, so it tends to cause the annular flow pattern. And vice versa, the higher the superficial liquid velocity, the smaller the void fraction value will be and make the condition irregular, causing the churn flow pattern.

Keywords: flow pattern, probability distribution function, two-phase flow, viscosity, void fraction.

1. Introduction

Void fraction is one of the essential physical values for gas-liquid two-phase flow in nuclear power plants, and it is an essential parameter for designs and performance evaluations of devices, including a core of BWR and so on[1]. Two-phase flow studies can be classified according to the position of the channel (horizontal, upright, straight, tilted) and according to the direction of the flow (opposite direction, horizontal direction, and vertical direction). According to Triplet et al. [2], the diameter of the channel is divided into 5, namely large channels, normal channel, mini channel, microchannel), and nanochannel. The two-phase flow in a mini channel is described in the absence of separate air and liquid flow. Sudarja (2018) states that the viscosity of the liquid affects the characteristics of the flow

pattern. In contrast to conventional channels, in mini-sized pipes, the surface tension is very influential on the two-phase flow characteristics, namely flow patterns, void fractions and pressure gradients.
Predicting the gas void fraction during two-phase flow in a channel is a crucial aspect of engineering calculations when designing process equipment. There are some well-known methods in the literature for calculating gas void fractions developed for two-phase flow in conventional channels. Unfortunately, there is still no method for calculating the gas void fraction that will apply to two-phase flow in a mini channel[3]. Research on two-phase flow and a void fraction has been carried out by previous researchers. Research has been conducted on small to large diameter channels, ranging from nanochannels, microchannels, normal channels, and large channels. The study produced different results because it used different methods. Sudarja et al. [4] have researched a void fraction of two-phase flow using a 1.6 mm diameter pipe using air and pure water with glycerin working fluids. Void fraction values are obtained by processing video images using a Nikon J4 camera with a speed of 1200 fps. This research was conducted in an adiabatic condition with a superficial gas velocity of 0.83 - 65.4 m / s and a superficial velocity of liquids from 0.02 to 4.14 m / s. The result showed that the flow pattern plays a vital role in predicting the void fraction. Whereas Sukamta et al. [5] conducted a study on the void fraction of two-phase airflow and a mixture of glycerin with different viscosities in capillary channels tilted 5° to the horizontal. Analysis of the results is done using digital image processing. The findings stated that the viscosity of the liquid is very influential on the bubbly and plug flow patterns. The increasing value of the gas superficial velocity causes the value of the void fraction to increase, and vice versa. Also, it is produced that the length of bubbly and plug flow patterns is influenced by homogeneous flow values (β).
Taisaku Gomyo and Hitoshi Asano [6] conducted a study of the characteristics of a void fraction in two-phase water-gas flow in pipes with small diameters. This research was conducted to determine the void fraction of the difference in a pipe diameter of 4 mm, 2 mm, 1.1 mm, and 0.5 mm. The void fraction was measured by the capacitance method and flow pattern using a high-speed camera. In this study, it was found that the wave frequency for annular flow increased when the diameter of the pipe narrowed to 1.1 mm. And for pipes with a diameter of 0.5 mm, the wave frequency decreases because of the increased surface tension effect.
Meanwhile, Pipathattakul et al.[7] conducted a study of the effects of 0°, 30°, and 60° angle variations on two-phase flow. This research was carried out on annular pipes with an inner diameter of 8 mm, 10 mm, and 11 mm and an outer diameter of 12.5 mm. The obtained flow patterns include plug flow, slug flow, slug-annular flow, annular flow, bubbly flow, churn flow, and dispersed bubbly flow. Based on this study, differences in angular variations affect different types of flow patterns.
Arun Autee et al. [8] conducted a study of pressure drop in two-phase flow by using angular variations of 30 °, 60 °, and 90 ° downward in pipes with diameters of 4 mm, 6 mm, and 8 mm and a channel length of 400 mm. The research uses a mixture of air and water as working fluids. The study was conducted to obtain the value of pressure drop and also compare the results with existing correlations of Crisholm on parameter C. It was found that the proposed correlation predicted a two-phase pressure drop at a satisfactory level. Hazard D. Baba et al. [8] examined the frequency of slug flow with high viscosity in two-phase flow. This research scale uses a mixture of oil and air as a working fluid and uses a pipe with a length of 17 m with a diameter of 76.2 mm. The results show that the increase in oil viscosity looks very influential on the frequency of the slug.
Ran kong [9] conducted a study on the effect of pipe diameter on bubble flow patterns in two-phase flow. The study was conducted using a 38.1 mm diameter pipe to the horizontal position. The working fluid used is a mixture of gas and air. From this study, it was found that an increase in the diameter of the pipe causes the bubble to be concentrated at the top of the flow. Chung and Kawaji [10] also took measurements of the void fraction. Different image analysis methods are used to distinguish the distance and average of the void fractions from the video images of gas and liquid interfaces. The void fraction data at 530 and 250 μm was reduced to reach 300 video images each time the experiment. A void fraction is obtained by matching the boundary occupied by the gas to symmetrical volume and estimating the fraction of the gas volume. This study produced a new slug flow model to predict physical insights about flow characteristics observed in microchannels. This model can predict two-phase friction pressure gradients for 100 and 50 μm diameters. Kawahara et al. [11] conducted a one-

phase and two-phase flow study with a working fluid, namely ionized nitrogen-water, using a 100 μm diameter fused silica channel. The parameters used are the air superficial velocity of 0.1–60 m / s and the superficial velocity of air 0.002–4 m / s. Flow patterns that were observed in this study were liquid alone (liquid slug) flow patterns, gas core flow with fine-thin liquid films, gas core flow with thick-fine liquid films, gas core flow with ring-shaped liquid films, and gas core flow with the deformed interface. A very small channel causes the Reynolds number to shrink and increase the surface tension that occurs so that the bubbly and churn flow in this study cannot be observed. Jiabin Jia et al. [12] performed calculations on the void fraction of two-phase flow at different pressures. The difference in pressure that occurs in the bubble and slug flow is substituted so that it gets a void fraction value. This research also discusses the effects of surface tension. The results show that friction loss cannot be ignored, especially when the void gas fraction is less than 0.2.
From the description above, it can be concluded that the void fraction is one of the parameters used to determine the characteristics of the two-phase flow. By determining the void fraction can be determined the type of flow pattern so that it can then be used to determine the high or low-pressure drop that occurs. By knowing the pressure drop, it can be used to predict the danger factors that will occur in a channel so that this information is vital to create an early warning system. Thus, research aimed at knowing the characteristics of the value of the void fraction in the two-phase flow, especially in the annular and churn flow, is essential

2. Method

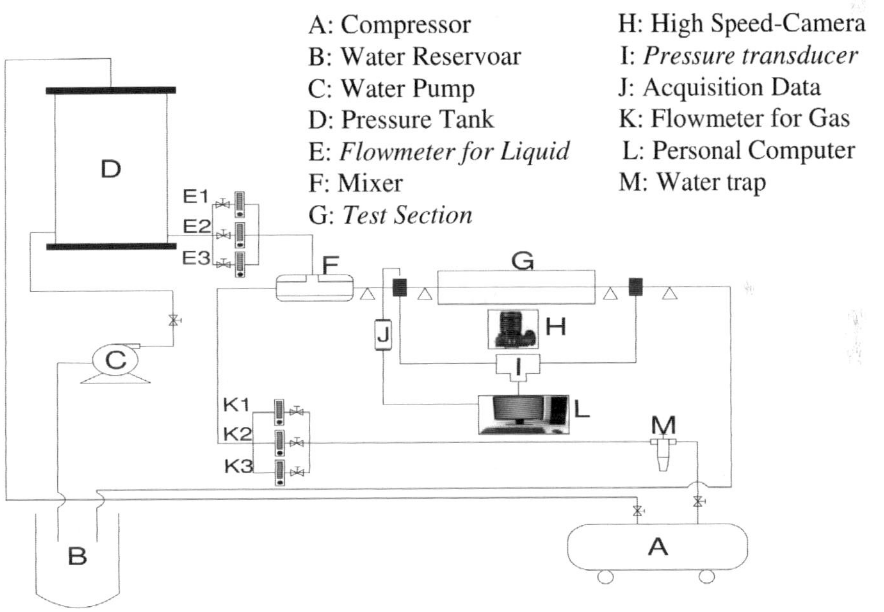

Figure 1. Scheme of Experimental Apparatus.

The study was conducted on an equipment installation shown in Figure 1. The equipment installation used in this study consisted of several main components: a water tank, a water pump, an air compressor, a pressure vessel, a mixer, a test section, a connector and a separator. Supporting equipment in this study include cameras, amplifiers, computers, optical correction boxes, acquisition systems, and video processing systems. Measuring instruments used in this study include Mpx, data acquisition, temperature indicator, pressure indicator, airflow meter, water flow meter. And thermocouple. Research material in the form of gas and liquid fluids. The gas fluid used is the air that has a low humidity obtained from a compressor that has been equipped with a water trap, while for the liquid fluid used is a mixture of water and glycerin with concentrations of 40%, 50%, 60% and 70%. This research was conducted at superficial gas velocity (JG) = 0.025 - 66.3 m / s and superficial velocity of fluid (JL) = 0.033 - 4.935 m / s. Data processing is used by Matlab and Microsoft Excel applications to get the Probability Distribution Function (PDF) value as a function of the void fraction value ε [-].

3. Results and Discussion

The annular pattern is formed when the superficial velocity of the gas (J_G) is relatively high, and when the superficial velocity of the fluid (J_L) is relatively lower. The high superficial velocity of gas causes the gas to penetrate the middle of the liquid film layer, and push the liquid film layer distributed evenly to the symmetrically at the pipe wall, as shown in Figure 2.

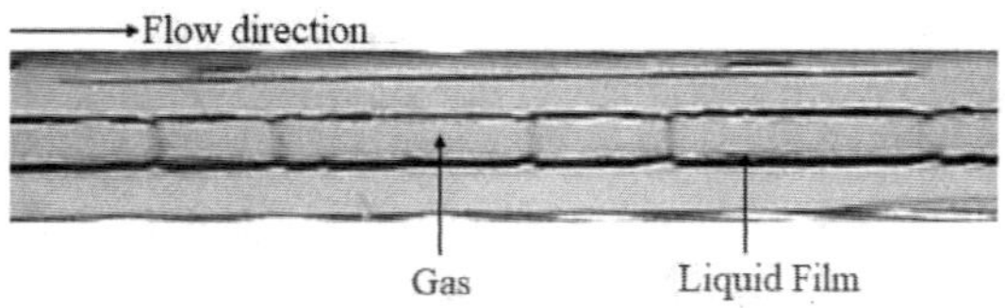

(a) Glycerin 40% and water 60%

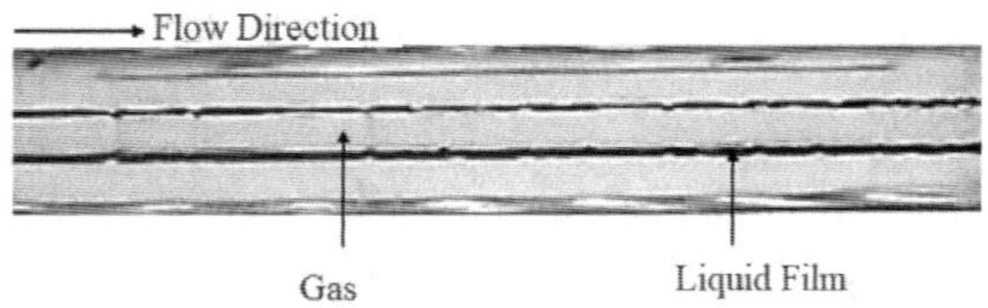

(b) Glycerin 50% and water 50%

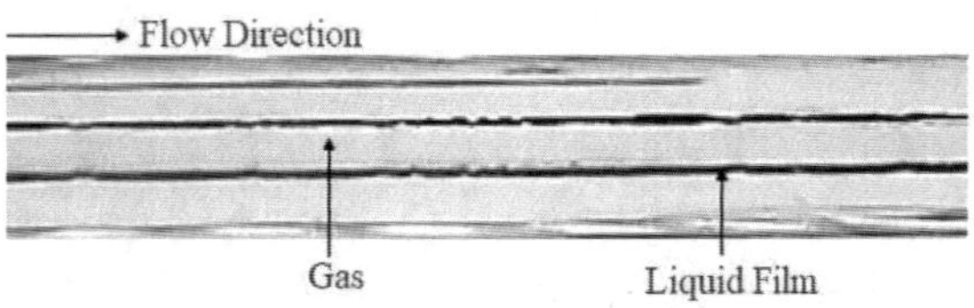

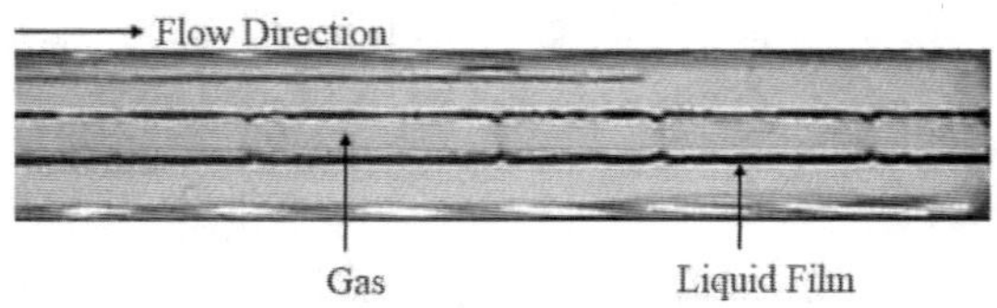

(c) Glycerin 60% and water 40% (d) Glycerin 70% and water 30%

Figure 2. Flow Pattern of annular (a) $J_G = 22.6$ m/s, $J_L = 0.149$ m/s, (b) $J_G = 50$ m/s $J_L = 0.149$ m/s, (c) $J_G = 50$ m/s $J_L = 0.149$ m/s, (d) $J_G = 50$ m/s $J_L = 0.149$ m/s.

Void fraction time series of annular flow pattern can be seen in Figure 3. From the figure, it can be an analysis that higher concentration of glycerin or higher viscosity of fluid causes an average of void fraction lower. This result means that the thick film layer on the wall is getting thicker so that it can increase the pressure drop.

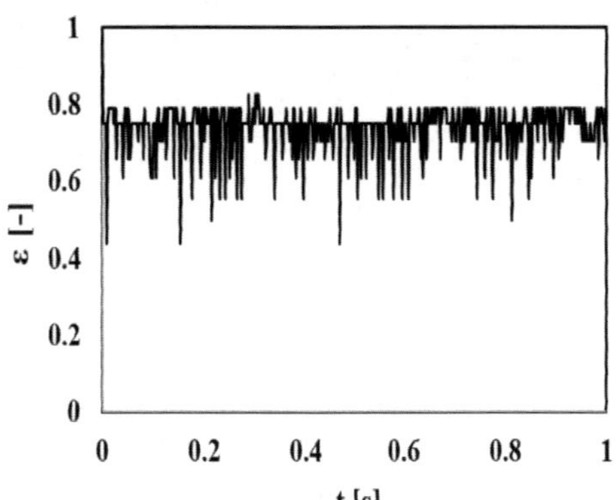

(a) Glycerin 40% and water 60%

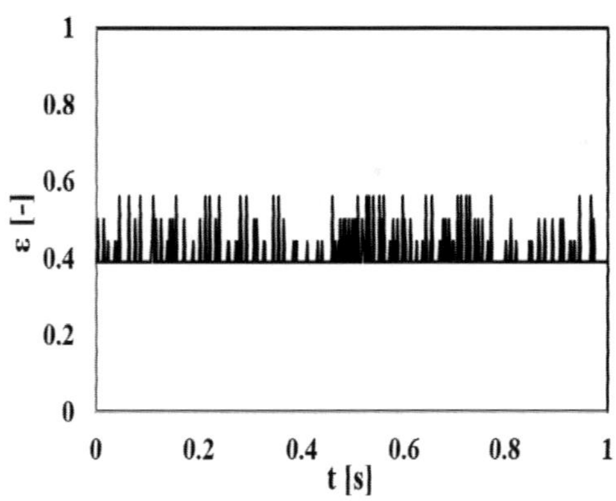

(b) Glycerin 50% and water 50%

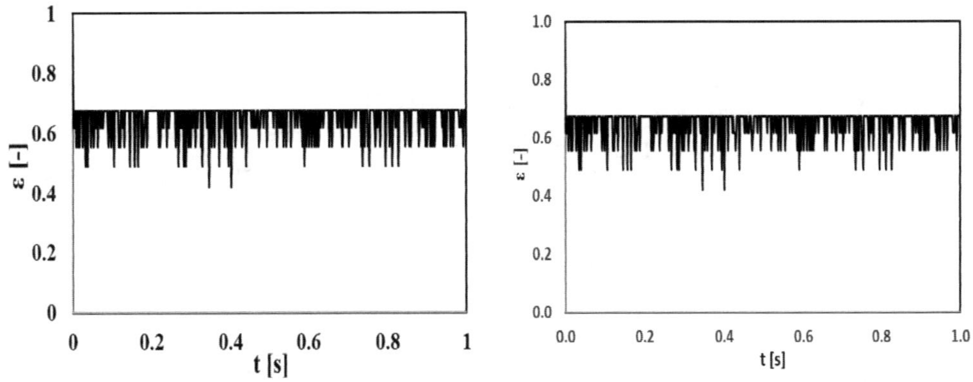

(c) Glycerin 60% and water 40% (d) Glycerin 70% and water 30%

Figure 3. Void Fraction Time series of Annular Flow Pattern (a) $J_G = 22.6$ m/s, $J_L = 0.149$ m/s, (b) $J_G = 50$ m/s $J_L = 0.149$ m/s, (c) $J_G = 50$ m/s $J_L = 0.149$ m/s, (d) $J_G = 50$ m/s $J_L = 0.149$ m/s.

This result is reinforced by Figure 4, which shows that as viscosity increases the peak value of the Probability Distribution Function shifts to smaller numbers. This result means that the void fraction is also getting lower.

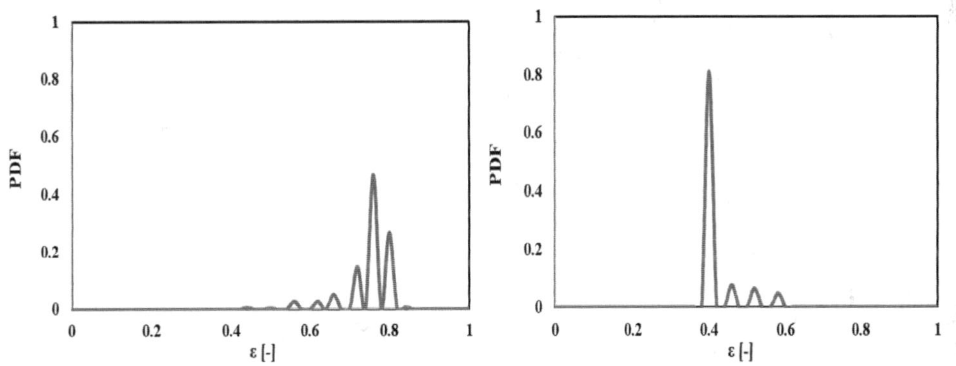

(a) Glycerin 40% and water 60% (b) Glycerin 50% and water 50%

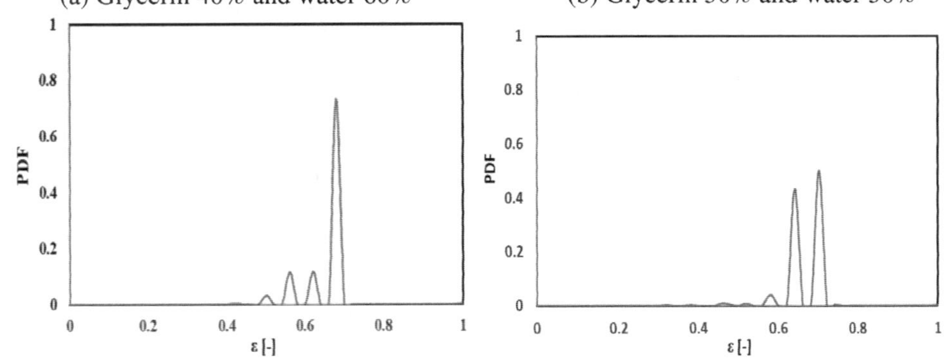

(c) Glycerin 60% and water 40% (d) Glycerin 70% and water 30%

Figure 4. Probability Distribution Function (PDF) (a) $J_G = 22.6$ m/s, $J_L = 0.149$ m/s, (b) $J_G = 50$ m/s $J_L = 0.149$ m/s, (c) $J_G = 50$ m/s $J_L = 0.149$ m/s, (d) $J_G = 50$ m/s $J_L = 0.149$ m/s.

In contrast, the churn flow pattern was shown in Figure 5. The churn flow pattern is formed when the superficial velocity of gas (J_G) is very high, and the superficial velocity of the fluid (J_L) is also very high, but both have a significant difference. The churn flow pattern is a continuation of the continued bubbly pattern but experiences instability in the back, resulting in a distorted part.

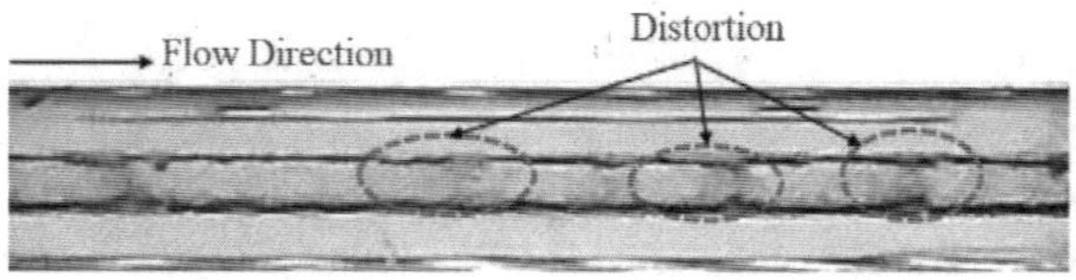

(a) Glycerin 40% and water 60%

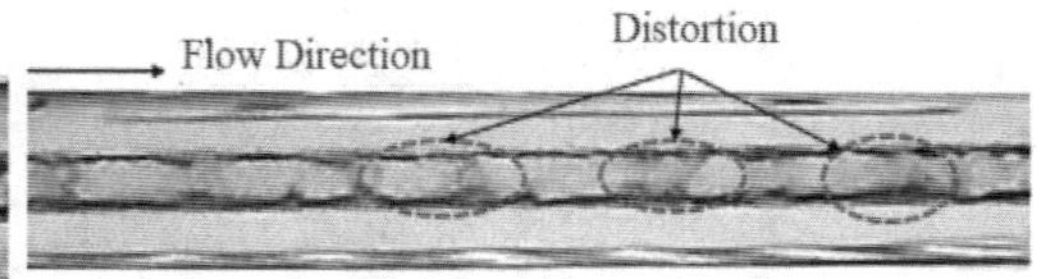

(b) Glycerin 50% and water 50%

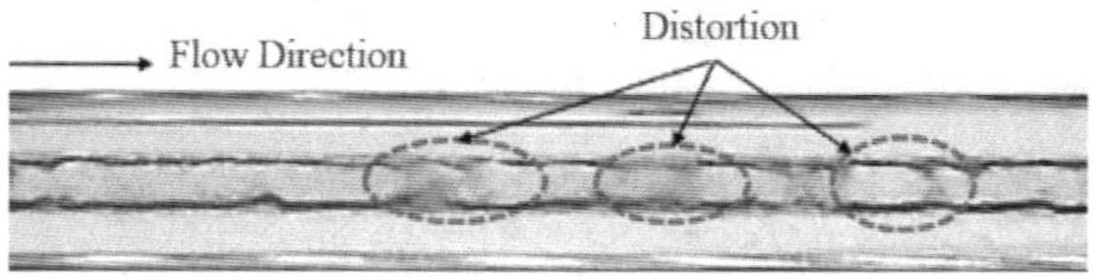

(c) Glycerin 60% and water 40%

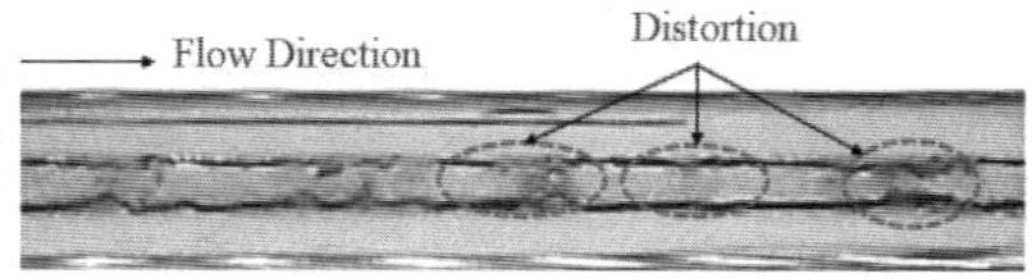

(d) Glycerin 70% and water 30%

Figure 5. Flow Pattern of *churn* (a) J_G = 58.05 m/s, J_L = 2.297 m/s, (b) J_G = 58.05 m/s J_L = 2.297 m/s, (c) J_G = 50 m/s J_L = 2.297 m/s, (d) J_G = 66.3 m/s J_L = 2.297 m/s.

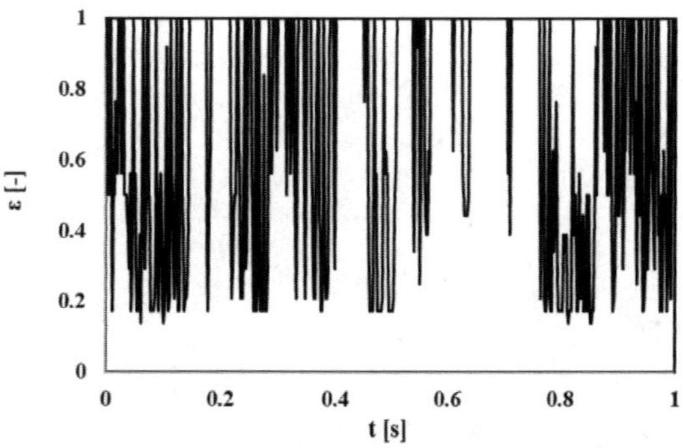

(a) Glycerin 40% and water 60%

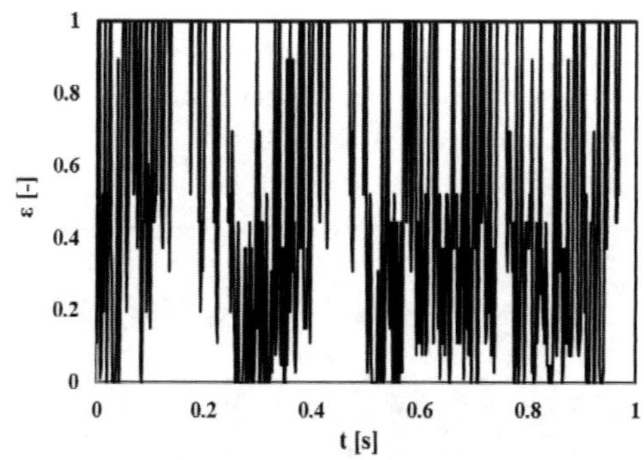

(b) Glycerin 50% and water 50%

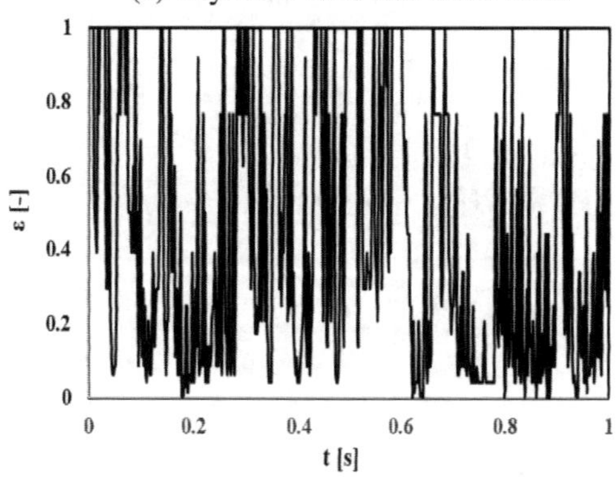

(c) Glycerin 60% and water 40%

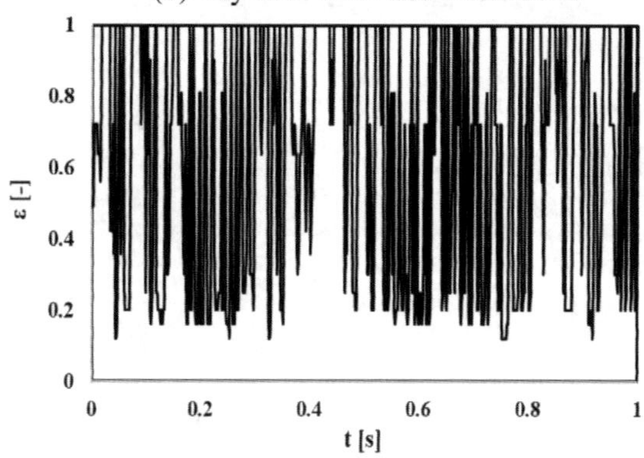

(d) Glycerin 70% and water 30%

Figure 6. Void Fraction Time Series of Churn Flow Pattern (a) J_G = 58,05 m/s, J_L = 2.297 m/s, (b) J_G = 58.05 m/s J_L = 2.297 m/s, (c) J_G = 50 m/s J_L = 2.297 m/s, (d) J_G = 66.3 m/s J_L = 2.297 m/s.

Figure 6 shows the void fraction time series of churn flow patterns from low to high superficial gas velocity with constant superficial liquid velocity. The results show that changes in gas superficial velocity affect the change in void fraction and not stable. Likewise, the increase in viscosity affects changes in the void fraction. This result is reinforced by the probability distribution function data, which tends to change with changes in the viscosity of the fluid as shown in Figure 7.

IOP Conf. Series: Materials Science and Engineering 1063 (2021) 012011 doi:10.1088/1757-899X/1063/1/012011

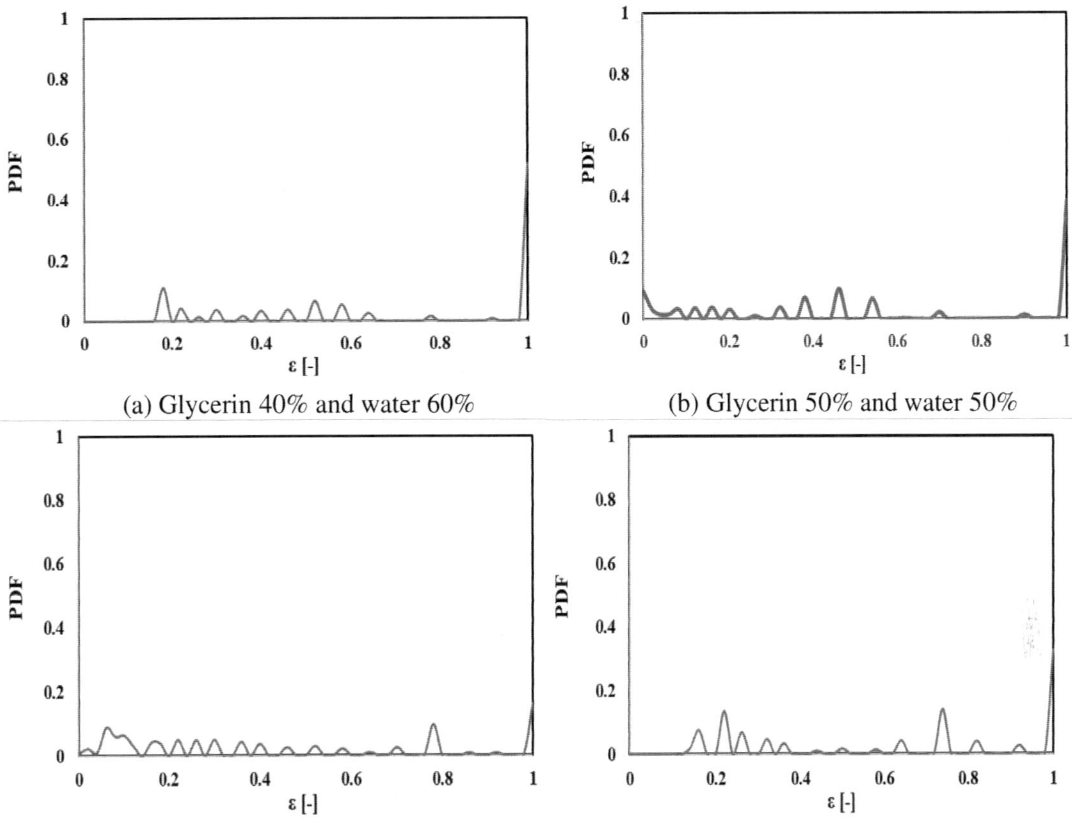

(a) Glycerin 40% and water 60%

(b) Glycerin 50% and water 50%

(c) Glycerin 60% and water 40%

(d) Glycerin 70% and water 30%

Figure 7. Probability Distribution Function (PDF) (a) J_G = 58.05 m/s, J_L = 2.297 m/s, (b) J_G = 58.05 m/s J_L = 2.297 m/s, (c) J_G = 50 m/s J_L = 2.297 m/s, (d) J_G = 66.3 m/s J_L = 2.297 m/s.

Meanwhile, the influence of the superficial velocity of gas and liquid can be seen in Figure 8. From the figure, it can be explained that the combination of the superficial velocity of gas and liquid is very significant in determining the type of flow pattern. For superficial velocity of gas, which tends to be high, and superficial liquid velocity, which is low, it forms an annular flow pattern. Vice versa, the superficial gas velocity which tends to be low and the high superficial liquid velocity form a churn flow pattern.

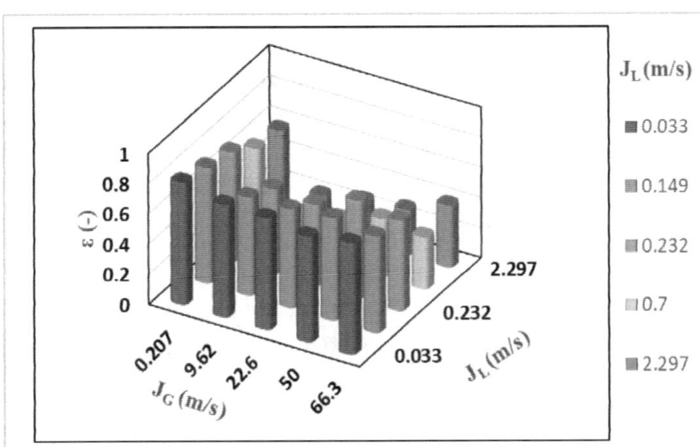

Figure 8. Effect of J_G and J_L to the Void Fraction

On the other hand, researchers have previously investigated the correlation of independent void fraction flow patterns for the flow of two gas-liquid phases in large vertical diameter annuli. They proposed two equations for the correlation parameters based on the drift flux model that involves the void fraction, fluid properties, pipe geometry, and phase flow rate. Experimental data are compared with literature and produce good accuracy[13]. Meanwhile, other researchers have also presented parametric analyzes of void fractions and then gave a brief review of some modeling techniques for determining void fractions and recommended recommendations for void fraction correlations that were well studied[14]. Previous researchers have also examined the characterization of void fractions. This researcher uses image processing on the temperature and pressure data flow of liquid nitrogen and gas to obtain the steam fraction and flow regime. This experiment is expected to lead to future research towards developing prototype void sensors for cryogenic two-phase flow[15]. Previous researchers have also experimentally investigated void fractions in the adiabatic two-phase flow of R1234yf in horizontal and smooth tubes with an internal diameter of 4.8 mm, vapor quality ranging from 0.1-1, saturation temperature of 15 and 25 ° C, and mass flow rate of 180 and 280 kg.m − 2s − 1. Volumetric fraction measurements using the fast-closing valve method. The result is that the empty fraction R1234yf is 5% lower than R134a[16]. Other researchers have reported the predictive performance of selected void fraction correlations based on three typical models under part for selected experimental conditions. The researcher furthermore shows that the correlation predicted based on flow patterns can predict the void fraction that is measured quite well[17].
Experimental data for void fraction and two-phase friction pressure reduction from various sources have been compiled and analyzed. Experimental data reveal that when superficial gas velocity and void fraction increase to higher values, the reduction in friction pressure becomes very sensitive to these two parameters. When the two-phase pressure drop is sensitive to variations in the void fraction, accurate characterization of the void fraction becomes significant[18]. Xing D et al. [19] reported that the correlation effect of the void fraction on the separation of pressure gradients from the two-phase flow of water in rectangular vertical channels. Superficial gas and liquid velocities range from 0.58 to 32 m / s and from 0.16 to 3.8 m / s. The results showed that although the void fraction correlation had a large influence on the gravity pressure gradient of slug flow, churn flow, and annular flow, and it had almost no influence on bubble flow. In the most common cases, many practical simulations make use of the "average void fraction of the system." The average void fraction of the system has also been estimated numerically based on heat transfer. A slip ratio based system means the void fraction is used to construct a moving boundary model for phase change heat exchangers. The transient startup prediction for the evaporator and condenser of the automotive air conditioning system is very much in accordance with the experimental results[20].

4. Conclusion

Based on the results and discussion, it can be concluded that the value of the void fraction in the annular flow looks more stable and does not show a significant change. The void fraction is also influenced by the level of film thickness in the annular flow. In contrast, the value of the void fraction in the churn flow has fluctuating values. At the same time, the effect of J_G and J_L on the value of the void fraction is very significant, where the higher the superficial gas velocity (J_G), the value of the void fraction will be higher, so it tends to cause the occurrence of annular flow patterns. And vice versa, the higher the superficial velocity of the liquid, the smaller the void fraction value will be and make the condition irregular, causing the churn flow pattern.

5. References

[1] S. Uesawa, A. Kaneko, Y. A.-F. measurement and instrumentation, and undefined 2012, "Measurement of void fraction in dispersed bubbly flow containing micro-bubbles with the constant electric current method," *Elsevier*, Accessed: Jul. 26, 2020. [Online]. Available: https://www.sciencedirect.com/science/article/pii/S0955598612000271.

[2] K. Triplett, S. Ghiaasiaan, … S. A.-K.-I. J. of, and undefined 1999, "Gas-liquid two-phase flow in microchannels Part I: two-phase flow patterns," *Elsevier*, Accessed: Jul. 25, 2019. [Online]. Available: https://www.sciencedirect.com/science/article/pii/S0301932298000548.

[3] M. Pietrzak, M. P.-I. J. of M. Flow, and undefined 2019, "Void fraction predictive methods in two-phase flow across a small diameter channel," *Elsevier*, Accessed: Jul. 26, 2020. [Online]. Available: https://www.sciencedirect.com/science/article/pii/S0301932219304665.

[4] Sudarja, Indarto, Deendarlianto, and A. Haq, "Experimental study on the void fraction of air-water two-phase flow in a horizontal circular mini channel," *AIP Conf. Proc.*, vol. 1737, 2016, doi: 10.1063/1.4949302.

[5] A. R. Sukamta and S. Ilham, "The Investigation Of Void Fraction Of Two-Phase Flow Air-Water And Glycerine (0-30%) In The Capillary Pipe With Slope Of 5 0 To Horizontal Position." Accessed: Jul. 26, 2020. [Online]. Available: http://journals.ums.ac.id/index.php/mesin/article/view/7385.

[6] T. Gomyo and H. Asano, "Void Fraction Characteristics Of One-Component Gas-Liquid Two-Phase Flow In Small Diameter Tubes," *Interfacial Phenom. Heat Transf.*, vol. 4, no. 1, pp. 1–18, 2016, doi: 10.1615/interfacphenomheattransfer.2016015746.

[7] M. Pipathattakul, O. Mahian, … A. D.-E. thermal and, and undefined 2014, "Effects of the gap size on the flow pattern maps in a mini-gap annular channel," *Elsevier*, Accessed: Jul. 26, 2020. [Online]. Available: https://www.sciencedirect.com/science/article/pii/S0894177714001472.

[8] A. Autee, S. S. Rao, R. Puli, and R. Shrivastava, "An Experimental Study on Two-Phase Pressure Drop in … THERMAL SCIENCE," *doiserbia.nb.rs*, vol. 19, no. 5, pp. 1791–1804, 2015, doi: 10.2298/TSCI130118081A.

[9] R. Kong, S. Kim, S. Bajorek, K. Tien, and C. Hoxie, "Effects of pipe size on horizontal two-phase flow: Flow regimes, pressure drop, two-phase flow parameters, and drift-flux analysis," *Exp. Therm. Fluid Sci.*, vol. 96, pp. 75–89, Sep. 2018, doi: 10.1016/j.expthermflusci.2018.02.030.

[10] P. M. Y. Chung and M. Kawaji, "The effect of channel diameter on adiabatic two-phase flow characteristics in microchannels," *Int. J. Multiph. Flow*, vol. 30, no. 7-8 SPEC. ISS., pp. 735–761, Jul. 2004, doi: 10.1016/j.ijmultiphaseflow.2004.05.002.

[11] A. Kawahara, P. Chung, M. K.-I. journal of multiphase, and undefined 2002, "Investigation of two-phase flow pattern, void fraction and pressure drop in a microchannel," *Elsevier*, Accessed: Jul. 25, 2019. [Online]. Available: https://www.sciencedirect.com/science/article/pii/S030193220200037X.

[12] J. Jia, A. Babatunde, and M. Wang, "This is a repository copy of Void Fraction Measurement of Gas-Liquid Two-Phase Flow from Differential Pressure. Void Fraction Measurement of Gas-Liquid Two-Phase Flow from Differential Pressure," *Elsevier*, 2015, doi: 10.1016/j.flowmeasinst.2014.10.010.

[13] N. Wang, B. Sun, P. Gong, and Z. Wang, "Improved Void Fraction Correlation for Two-Phase Flow in Large-Diameter Annuli," *Chem. Eng. Technol.*, vol. 40, no. 4, pp. 745–754, Apr. 2017, doi: 10.1002/ceat.201600119.

[14] A. J. Ghajar, "Void Fraction," in *SpringerBriefs in Applied Sciences and Technology*, Springer, 2020, pp. 37–64.

[15] G. Singh, S. Pradhan, V. T.- Cryogenics, and undefined 2019, "Experimental studies of two phase flow characteristics and void fraction predictions in steady state horizontal two-phase nitrogen flow," *Elsevier*, Accessed: Jul. 26, 2020. [Online]. Available: https://www.sciencedirect.com/science/article/pii/S0011227518302935.

[16] J. J. G. Pabon *et al.*, "Experimental study on the void fraction during two-phase flow of R1234yf in smooth horizontal tubes," *Int. J. Refrig.*, 2019, doi: 10.1016/j.ijrefrig.2019.05.007.

[17] Y. Zhou, H. Chang, Y. L.-E. T. and F. Science, and undefined 2019, "Gas-liquid two-phase flow in a horizontal channel under nonlinear oscillation: Flow regime, frictional pressure drop and void fraction," *Elsevier*, Accessed: Jul. 26, 2020. [Online]. Available: https://www.sciencedirect.com/science/article/pii/S0894177719306685.

[18] C. C. Tang, A. Ghajar, S. Tiwari, and A. J. Ghajar, "Effect of Void Fraction on Pressure Drop in Upward Vertical Two-Phase Gas-Liquid Pipe Flow," *asmedigitalcollection.asme.org*, 2012, doi: 10.1115/ICONE20-POWER2012-55167.

[19] D. Xing, C. Yan, X. Ma, L. S.-P. in N. Energy, and undefined 2014, "Effects of void fraction

correlations on pressure gradient separation of air–water two-phase flow in vertical mini rectangular ducts," *Elsevier*, Accessed: Jul. 26, 2020. [Online]. Available: https://www.sciencedirect.com/science/article/pii/S0149197013001601.
[20] S. P. Datta, R. K. Chandra, and P. K. Das Professor, "An Appraisal of System Mean Void Fraction and Its Application for the Moving Boundary Simulation of Phase-Change Heat Exchangers," *Artic. J. Heat Transf.*, 2015, doi: 10.1115/1.4030964.

Acknowledgment
Thanks to Silazyk Hafiz, who assisted in the measurement and processing of research data. We would also like to express our thanks to the Directorate of research and community service, the Ministry of Education and Culture, and the Muhammadiyah University of Yogyakarta for providing the budget for conducting research.

AUTHOR INDEX